Structures of agricultural education

Elements of the structure and terminology of agricultural education in India

Chandrika Prasad

The Unesco Press

Published in 1981 by the United Nations
Educational, Scientific and Cultural Organization
7 place de Fontenoy, 75700 Paris
Composed by Touraine-Compo, Chambray-lès-Tours
Printed by Offset Aubin, Poitiers

ISBN 92-3-101866-3

Preface

It is widely recognized that the development and utilization of human resources are key factors in social and economic progress, particularly among the rural population in the developing regions of the world. Thus, efforts to improve the qualitative standards and quantitative level of agricultural education and training in those regions are a necessity for technical advance in the agricultural sector, for increase in productivity, for changes in consumption patterns and, more generally, for the harmonious development of rural communities.

The World Conference on Agrarian Reform and Rural Development, organized by FAO in Rome (1979), and the World Conference on Agricultural Education and Training, held in Copenhagen (1970) and sponsored jointly by FAO, ILO and Unesco, have stressed the necessity of improving agricultural education systems and emphasized the need for more information and studies in relation to the problems of education for rural development. The diffusion and exchange of information on structures of agricultural education are considered essential factors for qualitative improvement of agricultural education.

Recognizing this need, Unesco has been publishing a series of detailed country studies on 'Structures of Agricultural Education'. The aim of these publications is the comparative study of agricultural education systems in various countries of the world. It is expected that the series will help agricultural educators, educational planners and administrators of agricultural education to improve the efficiency and output of agricultural education for rural development.

The present study, devoted to agricultural education in India, gives special attention to the relationship between agricultural education and general education, particularly the role of agricultural education institutions in non-formal education and agricultural extension. It describes the evolution of agricultural education in India before and after independence, defines the various levels of agricultural education: basic and multi-purpose schools, non-degree programmes, higher education. It also describes the pedagogy of teacher training in agriculture and the strategy of agricultural training for development.

Unesco wishes to thank Dr Chandrika Prasad, of the Indian Council of Agricultural Research, and the Indian National Commission for Unesco, under whose auspices the considerable work involved in the preparation of the present study was undertaken.

The author is responsible for the choice and the presentation of the facts contained in this book and for the opinions expressed therein, which are not necessarily those of Unesco and do not commit the Organization.

Contents

Other Unesco books on agricultural education:

Agriculture and the development process, by Louis Malassis.
Agricultural education in Asia. A regional survey.
Agriculture and general education.
Education in a rural environment.
Elements of the structure and terminology of agricultural education in Japan.

Introduction

Scope and importance of agricultural education and training

India lives in villages. There are 630,000 villages, comprising 80 per cent of the country's total population. Agriculture is the mainstay of the economy—nearly 70 per cent of the work-force is employed in agriculture and allied occupations. It contributes over 40 per cent of the national income, earns 42 per cent of foreign exchange and provides about 40 per cent of the raw materials to industry.

There are 70.5 million operational landholdings, of which 73 per cent are under 10 hectares in area. The average farm size in the country is 2.3 hectares, whereas the per capita arable holding is only 0.26 hectare. Against an overall literacy rate of 34.45 per cent, the rate in the rural areas is only 23.7 per cent (1971). While the population is increasing at a rate of 2.48 per cent, the overall growth rate in agriculture has been little over 3 per cent per annum in the recent past.

The implications of these basic facts in respect of agricultural education, research, training and extension can be readily appreciated.

Agricultural education in the Indian context is comprehensive: it encompasses crop technology, animal sciences, fisheries, agricultural engineering, forestry and rural home science, including crafts. Agricultural education also covers formal education from school to university level, as well as informal and non-formal education and training. It is a professional education—an instrument for bringing about desirable changes in rural structures, the economy and standards of living.

At the higher level of education, the teaching, research and extension functions are integrated.

Educational responsibilities and authorities

The Indian Constitution vests responsibility for 'education' primarily in the states, which function through their departments of education. The Ministry of Education and Social Welfare of the central government has been entrusted with responsibility for a few specialized fields of education and for some institutions. In addition, some of the states' projects are formulated and executed with the assistance of the central government. Recently, the constitution has been amended to make education, including technical, medical, and university education, a joint responsibility of the states and central government, subject to certain provisions.[1]

1. *Education in India, 1974-76*, New Delhi, Ministry of Education and Social Welfare.

The University Grants Commission (UGC) at central-government level is a statutory body which deals with university-level education. Several non-statutory and consultative bodies, such as the Central Advisory Board of Education (CABE), the National Council of Teachers' Education, the National Council for Women's Education, the National Council of Book Development Board, the National Board of Adult Education, the All India Council for Sports, the All India Council for Technical Education, the National Council of Educational Research and Training (NCERT), the Central Board of Secondary Education, etc., have been set up by government decision to carry out specialized tasks.

Responsibility for the supervision, direction and control of agriculture and animal science was entrusted to the states by some constitutional changes in 1919. The administration of the central agencies and institutions for research and for professional and technical training, however, was retained by the central government.

Agricultural education within the school system comes under the authority of the departments of education of the state governments. On the other hand, intermediate and vocational education in agriculture is the responsibility of the state departments of agriculture. While the management of agricultural universities and colleges is the responsibility of the state governments, they receive major grants-in-aid from the Indian Council of Agricultural Research (ICAR), an autonomous body set up under the Societies' Registration Act, 1860.

ICAR co-ordinates, aids and promotes research and higher education in agriculture and allied areas, but the co-ordination of agricultural development programmes, including extension training, is the responsibility of the Union Ministry of Agriculture and Irrigation. Nevertheless, ICAR has evolved innovative projects such as the National Demonstration, the Operational Research Project, the Krishi Vigyan Kendra (KVK) (Farm Science Centre) and teacher training centres (TTCs) as first-line demonstration projects and training institutions to serve as models for the rapid transfer of technology.

Structure of agricultural education and training

Pre-independence era

Agricultural education in school
In 1923 in Punjab province, an attempt was made to introduce elementary agriculture as an optional subject in the curriculum of vernacular middle schools involving both theoretical and practical instruction. Trained teachers (having taken the senior vernacular training course along with a course in agriculture) were provided to offer agricultural and allied courses in these schools. Either an agricultural farm (3 acres) or a school garden (1.5 to 1 acre) was used for the purpose of practical training.

The United province adopted the Punjab-type approach except that the agricultural course was compulsory for all boys from classes V to VII. In Bombay, schools having an agriculture bias were started and to a certain extent followed the Punjab pattern.

In 1937, the concept of 'basic education' was advocated by Mahatma Gandhi. This innovation in education was intended to bring about a radical change in the educational pattern—an educational system based on work and dignity of labour.

Students were involved in practical work through various village crafts, primarily agriculture.

By 1944, there were 269 basic schools in the country, including eight teacher training centres. On the re-establishment of the Congress government (1946), basic education was given considerable attention.

At the beginning of the present century, the state Department of Fisheries, Kerala, began operating fifty-four schools (three high schools, sixteen upper primary and thirty-five lower primary) with vocational fisheries as one of the subjects in the curriculum. At present, the Department of Education is responsible for these schools.

Non-degree educational programme
Soon after the reorganization of the states' departments of agriculture in 1905, the importance of agricultural education was recognized.

The first vocational school, known as the Marathi Agricultural School, was established in Bombay, at Kirkee, in 1910. This school aimed to train literate youths of the landholding class (at least fourth Marathi standard) in vocational agriculture and allied areas. When trained, students were expected to return to their farms and use their practical training to increase agricultural production. The course was of two years' duration leading to a certificate/diploma in agriculture. More such schools were established in the Bombay presidency (now Maharashtra state), Madras presidency (Tamil Nadu state), Central province (Madhya Pradesh state), United province (Uttar Pradesh state) and elsewhere.

The first veterinary school was opened in Poona in 1862 for training veterinary assistants and other field-level staff. A similar school was established in 1877 at Hapur, Uttar Pradesh. Civil veterinary departments were established in states in 1891 and created a demand for more trained personnel in this field.

In 1923/24, owing to the paucity of trained personnel in animal sciences, two diploma courses were launched: first, a two-year course leading to the Indian dairy diploma (IDD), and secondly, a fifteen-month course leading to the Associate IDD. The courses were initiated at the Imperial Dairy Research Institute, Bangalore (now the National Dairy Research Institute, Karnal), and a year later at Allahabad Agricultural Institute, Naini, Uttar Pradesh.

A course in home science at the intermediate level of the Board of Intermediate Education, Uttar Pradesh, was started in the Allahabad Agricultural Institute in 1943. Another course of a year's duration leading to a diploma in home economics extension was set up to train women extension workers in the same institute. The Lady Irwin College, established in 1932 at New Delhi, started a two-year diploma course in home science and a similar course for needlework.

The history of forestry education in India goes back to 1878, when a forestry school was established at Dehradun by the then North-western provinces and Oudh government (now Uttar Pradesh). This school was taken over by the Government of India in 1884 for the training of forest rangers. Another rangers' college was established in 1912 at Coimbatore, Madras.

Higher agricultural education
The importance of higher agricultural education was realized as early as 1870 when a few institutes were established to teach veterinary and agricultural sciences. In 1889, the Indian Veterinary Research Institute came into being at Poona, but was later moved to Mukteshwar, Kumaon (1893) and then to Izatnagar (1913).

The first five colleges of agriculture were set up at Kanpur (1906), Coimbatore (1906), Nagpur (1906), Poona (1907) and Sabour (1908). To start with they offered diploma/certificate courses. Later, degree-level education was introduced and the colleges were affiliated to general universities. A degree course in agricultural engineering was started in the early 1940s in the Allahabad Agricultural Institute.

A two-year postgraduate diploma known as an IARI Associateship was initiated in the Imperial Agricultural Research Institute, Pusa, in 1923. On the recommendation of the Royal Commission on Agriculture (1928), the Imperial Council of Agricultural Research was set up in 1929 to provide further encouragement and support to agricultural institutions. In the early 1930s postgraduate programmes leading to M.Sc. and Ph.D. degrees in agriculture were started. In the case of veterinary education, the Madras Veterinary College, in collaboration with the University of Madras, took the initiative of providing B.V.Sc. courses—a four-year degree programme—in 1936. Between 1946 and 1948, five further veterinary colleges were established at Mathura (1946), Rajendranagar (1946), Jabalpur (1948), Jorhat (1948) and Hissar (1948).

By 1947, the year of independence, there were seventeen institutions offering higher degrees in agriculture and having an annual enrolment of about 1,500 students. The research and extension functions at that time remained the responsibility of the state departments of agriculture.

Teacher training in agriculture
During the pre-independence period, teacher training in agriculture was almost non-existent. Nevertheless, a few instances have been noted where some orientation was given to teachers in agricultural schools by some of the agricultural institutions. One of the oldest agricultural schools, at Bulandsahr (1921) in the United province, had a deliberate policy for the training of teachers of the senior basic schools.

A one-year teacher training diploma was offered to the two-year diploma holders in home science in the early 1930s. The Vidya Mandir Training School was established in Wardha, Maharashtra, in 1938 as the first basic school for teachers.

Agricultural training for development
Prior to independence, rural development did not receive adequate attention. The training programmes were limited mainly to administrative aspects and focused on bureaucratic methods and approach. Training for development was a rarity.

In the early 1920s, efforts were made, here and there, both by the government and by voluntary organizations, to ameliorate the condition of rural people, particularly through agricultural development. The Gurgaon project (1927), the Martandam project (1921), the Shriniketan project (1920), etc. provide examples. Some training efforts, both for extension/social workers and rural people, were attempted by them in a more or less sporadic and casual manner. A deliberate effort, however, was made in the Shriniketan project, where young farmers in particular were brought to the projects' scientifically managed farms, and after training were expected to cultivate their own farms on scientific lines.

At Shriniketan, a boarding school—Siksha Satra—was established for the training of young farmers. A selected group of boys was brought to the school every Monday and after a week of training returned to their villages on Saturday. They brought with them rice, etc. for meals during their one-week stay. As part of their training, they were assigned to a variety of activities in poultry and dairy farms, the school garden, kitchen and dormitory. Besides agriculture and animal husbandry, they were taught home crafts, songs, plays, etc. In the Martandam project, the training was focused on the village workers. The first course (of ten days' duration) was offered in 1926 at Quilon. Subsequently, a training school was built from bamboo poles and coconut leaf thatching at a total cost of only Rs.21. Eventually, a full one-year training course was organized for rural reconstruction workers.

Agricultural development began to receive more attention from the government following the report of the Royal Commission on Agriculture. The commission recommended that training should be provided for farmers as well as in-service training for development staff in agriculture and allied areas. Consequently, some training programmes were organized for farmers, mainly on government farms, and agricultural shows/exhibitions were utilized as a medium of training.

Short training courses of two to three months' duration were organized in animal husbandry and dairy farming at the cattle breeding farm, Karnal, Haryana, the Indian Agricultural Research Institute, Pusa, Bihar, the Central Creamery, Anand, Gujarat and the Dairy Institute, Bangalore.

The need for fisheries training was felt early in this century. At that time, training for fisheries operatives and others was given on an ad hoc basis on the exploratory fishing vessels of the Government of India, manned by personnel trained in the United Kingdom. As part of the post-war reconstruction programme of the country in the 1940s, fisheries training was considered a multi-disciplinary subject and localized training programmes were started by provinces such as Madras and Bombay.

Post-independence era

Agricultural education in schools
The Indian Constitution provides for free and compulsory education for the 6–14 age-group (classes I–VIII). In pursuance of this constitutional directive, free education has been introduced in primary schools in all states for the 6–11 age-group (classes I–V). Free education for the 11–14 age-group has also been introduced in all states except Orissa, Uttar Pradesh and West Bengal. In 1977/78, 67 per cent of the 6–14 age-group (81.7 per cent boys and 51.4 per cent girls) were availing themselves of the educational opportunities in schools. Recently, some states have also decided to provide free education up to secondary level for the 15–17 age-group.

In the early post-independence period, basic education was enthusiastically supported and reinforced. On the recommendation of the Education Commission (1964–66), a large number of middle schools in Uttar Pradesh were oriented as senior basic schools by introducing subjects such as agriculture, woodcrafts, tailoring, metal crafts, spinning and weaving, leather crafts and home crafts. Agriculture as a main craft was introduced in 52,654 junior schools (classes I–V) and in 2,538 senior basic schools (classes VI–VIII), especially in rural areas. While there were no provisions for separate agricultural teachers in junior basic schools, such teachers, duly oriented and trained, were provided in senior basic schools.

Although this movement took place in other states also, it did not make much headway. The conversion of most of the schools into basic schools proved difficult and an effort was therefore made to introduce at least a few of the salient features of the basic school curriculum in the school system in general. In 1956, the National Institute of Basic Education was established to conduct research on the subject and to provide extension services to support basic education.

The Mudaliar Commission (1952/53) recommended the establishment of multi-purpose high/higher secondary schools where, in addition to core subjects, one group of subjects, namely, humanities, sciences, agriculture, commerce, technical subject, fine arts and home science, was to be taught. Some of the general secondary schools also offered agriculture as a subject. In some cases it was an elective subject, while in others it was substituted for one of the crafts. Technical high schools (mostly in Maharashtra and Gujarat states) offered technical subjects, including agriculture, in classes IX and X. In Gujarat state there were a few privately run post-basic schools with courses in agriculture.

Tables 1 and 2 show the number of schools, enrolment, and number of teachers in both general and vocational schools.

TABLE 1. Number of general schools, enrolments and teachers (1972/73)

Schools	Number of schools	Enrolment		Total	Number of teachers	Average number of students per teacher
		Boys	Girls			
High/higher secondary schools	39 764	11 806 101	5 213 378	17 019 383	694 865	25
Middle/senior basic schools	97 623	13 948 770	7 733 847	21 682 617	692 263	31
Primary/junior basic schools	431 791	28 403 474	16 872 230	45 275 704	1 150 805	39
Pre-primary/ pre-basic	4 806	183 512	168 621	347 133	9 555	36

Source: Education in India, 1972-73, New Delhi, Ministry of Education and Social Welfare, Government of India.

The concept of basic education, as well as the introduction of multi-purpose schools and some of their variants, met with only limited success. The intricacy of the subject could not be fully appreciated and translated into reality for various reasons. In view of the discouraging results, the Education Commission suggested that this system be discontinued and recommended that all primary schools, in both rural and urban areas, should give an agricultural orientation to their existing courses in general science, biology, mathematics, social sciences and other disciplines. The National Commission on Agriculture (1976) also endorsed this view and further suggested a revision of general science textbooks, especially biology, so as to orient them towards agriculture.

TABLE 2. Number of vocational/professional schools, enrolment and teachers (1972/73)

Vocational/ professional schools	Number of schools	Enrolment		Total	Teachers (total)	Average number of students per teacher
		Boys	Girls			
Agriculture and forestry	102[1]	5 268	125	5 393		
Technical and industrial education	1 625	64 151	42 575	106 726		
Compounding	3	171	8	179		
Nursery and midwifery	101	11	5 175	5 186	1 377	13
Physical education	18	2 216	264	2 840		
Teacher training	198	9 730	6 746	18 476		
Others	374	19 630	16 380	36 010		

1. Includes eighteen agricultural schools, in Uttar Pradesh also.
Source: Education in India, 1972-73, op. cit.

Following the recommendations of the Education Commission (1964–66), the government has accepted the 10 + 2 + 3 pattern of education. This pattern provides for work experience to be given to students during their ten years of schooling and, at the higher secondary stage, vocational courses to be introduced in addition to the core subjects. Given its importance, agriculture has priority in the scheme of work experience as well as in the vocational courses.

Since 1947, eight fisheries primary schools in the coastal districts of Thane, Kulaba and Ratnagiri in Maharashtra state offer a course on vocational fisheries as part of the school curriculum. Over 250 school-going youngsters have access to such training. From 1964, the State Fisheries Department has been running a fisheries high school in Thane. There are eleven high schools and eighteen Zilla Parishad schools in the private sector that offer fisheries as one of the subjects. Since 1967, three regional technical high schools specializing in fishery have been established in Kerala state. Every year forty students are admitted at the eighth grade and are taught fisheries technique through the tenth grade.

Non-degree educational programmes
Vocational education in agriculture and related areas continues to be offered through agricultural and allied schools at certificate and diploma levels, but with the establishment and extension of higher educational institutions offering degree programmes, the number of diploma/certificate courses and institutions has been relatively reduced.

The Community Development Programme was launched in 1952 for integrated rural development. This project required a large number of multi-purpose village-level workers (VLWs) both male *(gram sevak)* and female *(gram sevika)*.

Initially, five training-cum-development units were established and subsequently forty-three extension training centres. A six-month composite training course was organized for VLWs in the fields of agriculture, livestock production, co-operation, public health and hygiene, social education, extension methods and approach, etc. This certificate course was not completely adequate, and in 1953/54 basic agricultural schools were established with training programmes of one and half years consisting of one-year agricultural courses followed by six months of training in extension methods and such subjects as co-operation, public health and minor engineering.

Since 1958, the training programmes for VLWs have been further improved by the introduction of a two-year integrated diploma course with considerable emphasis on agricultural subjects (one and a half years in agricultural school and six months in an extension training centre). The same course was later extended to two and a half years (two years in agricultural school and six months in an extension training centre). One hundred such extension training centres were established throughout the country.

In 1955, extension activities for rural women were undertaken as an integral part of the Community Development Programme (CDP). To train *gram sevikas*, home science units were attached to the extension training centres to provide a one-year pre-service certificate course. In order to cope with the number of *gram sevikas* required, forty-six such *gram sevikas* training centres (GTC) were opened.

After independence, a rapid expansion of the development programme under the successive five-year plans required large numbers of trained personnel at professional and subprofessional levels. One- or two-year certificate/diploma courses were offered in the fields of animal husbandry, veterinary science and dairy farming, farm technology (including agricultural engineering), home science, fisheries and forestry, etc. to meet the increasing demand for trained personnel.

There was also a shortage of women especially trained for rural home management. Initially, most of the home science colleges of the agricultural universities offered non-degree programmes. For instance, in 1966 the Punjab Agricultural University, Ludhiana, offered a one-year certificate course for farm women. Since 1971, the Gobind Ballabh Pant University of Agriculture and Technology has offered a two-year diploma in home science. The Gujarat Agricultural University has been running a home science school at Anand, which offers a two-year diploma course. The School of Baking at Anand has offered a five-month certificate course in bakery since 1963.

The Agricultural Engineering Institute, Raichur, introduced a diploma course in agricultural engineering in 1970, with two specializations—farm machinery and soil and water management. In order to strengthen forest ranger training programmes, two additional rangers' colleges—Eastern Forest Rangers College, Kurseong, West Bengal, and Central Forest Rangers College, Chandrapur, Maharashtra—were established in 1974 and 1976 respectively. The IDD (dairy technology) and IDD (dairy husbandry) courses, initiated in the early 1920s, were continued.

The training of veterinary fieldmen/stockmen[1] has been organized through one-year certificate courses that are being offered in a dozen institutions. A one-year certificate course has also been organized for veterinary compounders[2] in several institutions.

___

1. Veterinary workers engaged in gelding cattle.
2. Qualified (first-level) workers in veterinary dispensaries.

Based on the Shrimali's Committee Report, the Union Ministry of Education established rural institutes (1956) in association with already existing educational institutions/agencies. Fourteen rural institutions were established in different parts of the country which initially offered one- to three-year diploma courses in such subjects as agriculture, technical activities, sanitation, general education, etc. The basic aim of the institutes was to give education relevant to the rural masses.

Fisheries education and training have only been given due importance in the recent past. Some training courses in inland and marine fisheries were started at the Central Inland Fisheries Research Institute (CIFRI), Barrackpore, and the Central Marine Fisheries Research Institute (CMFRI), Madras, between 1948 and 1950. A few polytechnics were also established in the states of Kerala, Andhra Pradesh and Tamil Nadu.

The first centre for training fishery operatives was established in 1955 at Satpathi in Maharashtra state. Since then over forty centres have come into being for strengthening such training programmes. Facilities for the training of personnel, e.g. skippers and engineers of fishing vessels, were made available in 1948 through the Deep Sea Fishing Organization, Bombay. The Government of India opened the Central Institute of Fisheries, Nautical and Engineering Training at Cochin in 1963 for training skilled personnel for manning shore-based establishments. A branch of this institute was opened at Madras in 1968.

On the recommendations of the Committee on Fisheries Education (1959), the Central Inland Fisheries Education (CIFE) and Central Inland Fisheries Operatives (CIFO) were established in 1961 and 1963 at Bombay and Cochin respectively. Two regional training centres for inland fisheries operatives were opened in 1967 at Agra and Hyderabad. Later, CIFRI was transferred to ICAR from the Ministry of Agriculture and the training programmes organized at the regional training centres were taken up by CIFE. In 1973, the regional training centre in Hyderabad was converted into an extension training centre for senior officers in inland fisheries.

Figure 1 shows the educational system of the country.

Higher agricultural education
The University Education Commission (1949), headed by Dr S. Radhakrishnan, recommended the setting up of 'rural universities' on the pattern of the land-grant colleges of the United States. The first joint Indo-American Team (1955), after studying the existing research and educational infrastructure in agriculture, endorsed the recommendation of the University Education Commission and suggested the creation of 'agricultural universities'. Towards this end, a twinning relationship between five land-grant institutions of the United States and the agricultural institutions in India was developed for technical back-stopping. The first agricultural university was established in 1960 at Pand Nagar, Uttar Pradesh (now known as Gobind Ballabh Pant University of Agriculture and Technology). The concept of an agricultural university was a significant innovation—a turning-point in the history of agricultural education in the country.

The second joint Indo-American Team (1959) assessed the progress made in implementing the earlier recommendations on agricultural education and gave necessary directions for the Third Five-year Plan (1961-66). One of the recommendations was to develop a well-defined agricultural education system in the country comprising higher as well as lower agricultural education (agricultural education in the schools). In order to provide strong support to agricultural institutions,

Elements of the structure and terminology of agricultural education in India

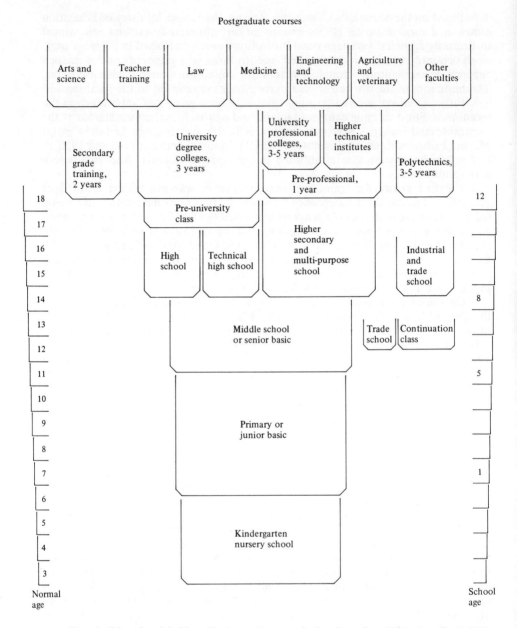

Fig. 1. Educational ladder. (Source: *Report of the Secondary Education Commission* (1952-53).)

ICAR was reorganized on the basis of the recommendation of the Agricultural Research Review Team (1963) composed of experts from the United States, the United Kingdom and India (Fig. 2).

The Education Commission (1964–66) suggested the setting up of one agricultural university in each state. At present, all the states have established agricultural universities (numbering twenty-one), except Jammu and Kashmir and the small states of the North-eastern hill region. The Education Commission also noted that in the past greater emphasis had been placed on higher education as compared to agricultural education at lower levels.

In addition to the constituent colleges of the agricultural universities, there are thirty-four agricultural colleges affiliated to general universities. Some of the general universities also offer postgraduate degree programmes in applied biological sciences—agricultural botany, agricultural zoology, agricultural chemistry, etc.

Out of thirty-one institutes controlled by ICAR, four institutes, namely the Indian Agricultural Research Institute (IARI) (1923), the Indian Veterinary Research Institute (IVRI) (1913), the National Dairy Research Institute (NDRI) (1955) and the Institute of Agricultural Research Statistics (IARS) (1959), provide postgraduate degree/diploma programmes in their respective disciplines. Such Indian agricultural research institutes are 'deemed' to be universities, and other institutes have been recognized by the general universities for the award of postgraduate degrees/diplomas.

There are four institutes of management outside the purview of the agricultural universities that offer postgraduate diplomas in business management and administration as applied to agriculture. Since 1971, the Punjab Agricultural University (PAU), Ludhiana, has initiated an M.B.A. degree programme in agriculture.

Fisheries education at degree level was only introduced in 1969 at the College of Fisheries at Mangalore University of Agricultural Sciences, Bangalore, Karnataka. A similar course of B.F.Sc. is given at CIFE, Bombay. Two-year and one-year postgraduate courses in both inland and marine fisheries are offered at CIFE and the Inland Fisheries Training Unit (IFTU) to train fisheries development staff for the state governments.

A fifteen-month postgraduate diploma in sericulture is offered at the Central Sericulture Research and Training Institute, Mysore, Karnataka. Education and training in forestry at degree level is provided at the Indian Forest College, Dehradun, Uttar Pradesh, and the State Forest Service College, Burnihat.

Teacher training in agriculture
The first session of the Indian Council of Agricultural Education (ICAE), 1952, which was held at Hyderabad, recommended the organization of schoolteacher training in agricultural schools and colleges. The second session of ICAE, held in 1956 at Lucknow, suggested the organization of special training conferences for veterinary teachers. The joint Indo-American Team on agricultural education, research and extension (1959) recommended that the suggestion of ICAE to conduct teaching seminars and conferences should be followed up. The team further recommended that 'there should be established teacher training departments in conjunction with agricultural colleges'.

The Central Institute of Fisheries Operatives (1963) offered a teacher training course awarding certificates to successful candidates. The Education Commission (1964-66) recommended the reorganization of in-service training for teachers in the

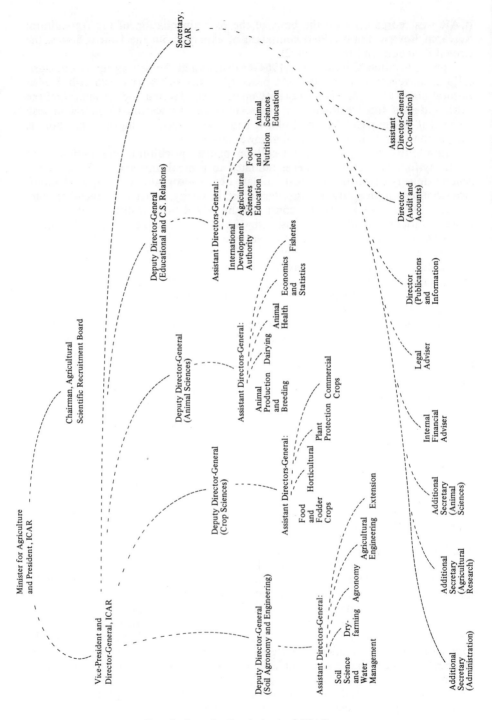

Fɪɢ. 2. Organizational chart of ICAR.

form of summer institutes and specialized courses by the agricultural universities. Accepting the recommendations, ICAR formulated and implemented a scheme for summer institutes in 1965.

ICAR had also organized a teacher training workshop/seminar at Trivandrum, Kerala, in 1957; at Mussouri, Uttar Pradesh, in 1958; and at Bombay, Maharashtra, in 1960. Since then, agricultural universities/colleges have been periodically entrusted with the conduct of such workshops in their respective institutions.

During the Third Five-year Plan (1961-66), four regional colleges of education were established at Ajmer, Bhopal, Mysore and Bhubaneshwar for the purpose of effecting qualitative improvement in school teaching staff by the National Council of Educational Research and Training. The integrated training included general education, specialized subject-matter areas and teaching methodology (B.Sc. in education, M.Sc. in education, physics, etc.). One- to four-year courses were offered in various disciplines including agriculture. Extension service departments were also established in fifty-four selected teacher training colleges for in-service training.

In the initial phase of the Community Development Programme (CDP), the training of the teachers, principals and instructors of *gram sevak* and *gram sevika* training centres was organized in the form of short-term workshops. Recognizing the continuing nature of such training, an extension education institute was established in 1959 at Nilokheri, Haryana. In 1962 two more such institutes were opened, one at the Osmania University (now Andhra Pradesh Agricultural University), Hyderabad, and another at the Institute of Agriculture (Gujarat Agricultural University) at Anand.

Recognizing the importance of teacher training in agriculture, the Punjab Agricultural University, Ludhiana, initiated a postgraduate course, M.Sc. (agricultural education) in agriculture in 1971. One-year teacher training courses in agriculture, home science and agro-mechanics have recently been formulated by Punjab Agricultural University. The course with respect to home science began with the academic year 1977/78. Another in-service teacher training programme of one year's duration is specially oriented to specific subject-matters. Occasionally, ad hoc in-service courses for teachers are also provided.

Since 1976, TTCs in specialized disciplines—dryland agriculture, horticulture, fisheries, dairy farming, home science, etc. have been established by ICAR, mainly at its specialized research institutes. These centres intend to organize skill-oriented in-service training courses for the trainers/teachers involved in non-degree education programmes and training.

Agricultural training for development
Education and training are life-long requirements. This is particularly so in the present era when scientific advances must be tapped in order to improve the life and living standards of ever larger populations. People at all levels—professional, subprofessional and production—must keep abreast of the latest developments in order to take full advantage of the scientific breakthrough in agriculture and allied fields. A strong induction and in-service training mechanism is thus an essential element for raising agricultural and allied production.

The second joint Indo-American Team (1960), the Education Commission (1964–66) and the National Commission on Agriculture (1976) emphasized the necessity for strengthening the training infrastructure in order to raise the professional and technical competence of extension workers. Special units for refresher training programmes of the *gram sevaks, gram sevikas,* and *mukhya sevikas* were created in selected *gram sevaka* training centres. The training courses offered were of varied durations—two months to one year depending on specific local needs. These centres also organized ad hoc training courses for farmers, farm women and young boys and girls.

The Directorate of Extension Education and departments of agricultural extension of agricultural universities/colleges have also been organizing in-service training courses for the extension staff of the respective state governments, and the farmers and farm youths of villages in the vicinity of their campuses.

The training of higher level managerial and administrative staff is equally important. The National Commission on Agriculture has placed particular stress on this aspect. In order to meet this need, ICAR started the Central Staff College for Agriculture at Hyderabad in 1976. The main aim of this college is to organize staff courses, seminars and conferences on agricultural management and administration of research, education and development systems and institutions. Short-term courses for the Agricultural Research Service of the ICAR probationers have also been organized at the staff college, with the emphasis on the total research, education and development responsibility of the council, the institutional infrastructure which exists in the country and the collaborative and interdisciplinary efforts required in modernizing agriculture.

In view of the intensification of agricultural development programmes through community development and various schemes such as the Intensive Agricultural District Programme (IADP) in 1960-61, the Intensive Agricultural Area Programme (IAAP) in 1964, the Small Farmers Development Agency (SFDA), the Applied Nutrition Programme (ANP), the Drought Prone Area Programme (DPAP), etc., the vital roles of the agricultural universities and colleges and of the agricultural research institutes in training personnel for development have been fully realized. These institutions are playing a most useful role in the field of training, especially for state and district-level extension personnel.

In view of the massive need for training of farmers, farm women, fishermen and school drop-outs, the ad hoc training programme organized by various institutions was considered grossly inadequate. Thus, a farmers' training and education programme was launched in 1959 by the Directorate of Extension, Ministry of Agriculture, Government of India. There are now 150 farmer training centres in different parts of the country, especially in areas of intensive development.

An innovative training institution called Krishi Vigyan Kendra (KVR) (Farm Science Centre) was evolved and established by ICAR in early 1974 at Pondicherry under the supervision and administrative control of Tamil Nadu Agricultural University (NAU), Coimbatore. This institution aims at imparting skill-oriented training to practising farmers, farm women, school drop-outs, illiterate farm youths (boys and girls), agricultural labourers and field-level extension staff. The training is provided through work experience on the principle of learning by doing. During 1976-78, a further nineteen Krishi Vigyan Kendras were established in sixteen states or union territories.

Short training courses in animal sciences, on both a regular and an ad hoc basis, are organized in most of the animal science colleges/institutes/departments. For example, the southern regional station of the National Dairy Research Institute (NDRI) at Bangalore offers a three-month training course on a regular basis starting in December each year and organizes special short courses of varying duration as the demand arises.

22

Short-term courses in fish handling and freezing are offered through the Integrated Fisheries Project under the Ministry of Agriculture and Irrigation and the Central Institute of Fish Technology (ICAR), Cochin.

Since 1974, the Central Institute of Fisheries Operatives has offered short-term training courses (seven to ten days) for personnel on fishing vessels, skippers, engineers, etc. Courses designed for extension workers have also started in several institutions.

Refresher training courses in sericulture have been made available at the Central Sericulture Research Institute, Mysore.

Many more field-level training courses are now being organized by state governments and some of the voluntary institutions committed to rural development.

Agricultural training through non-formal programmes
The Ministry of Education (Department of Social Welfare) and the Ministry of Agriculture and Irrigation (Department of Rural Development) are both engaged in rural regeneration through education and development. Several of their social educational programmes include elements of agricultural training, especially for rural and urban youth.

The National Cadet Corps was organized in 1948 to train young men and women. The Planning Forum movement was started in 1955 in order to develop planning consciousness among university/college teachers and students as well as to associate them with the formulation and implementation of the five-year plans. Since 1959 the concept of rural youth clubs *(yuvak mandal)* has been launched as a part of the Community Development Programme (CDP). The Radio Rural Forum (Charcha Mandal) was also initiated in 1959.

The Farmers Functional Literacy Project (Kisan Saksharta Yojna) was introduced in 1967–68 with an organic linkage with the farmers' training programmes. The National Service Scheme (NSS) was begun (1969–70) in the higher educational institutions in order to involve staff and students in social services, mainly in the rural areas. The Non-formal Education Project, which started in 1974-75, stressed non-formal education for non-school children in the 6-14 age-group, for youth in the 15-25 age-group and the linking of functional literacy with development programmes.

On the basis of the recommendation of the National Advisory Board on Youth (1970), a scheme for the establishment of the Nehru Yuvak Kendra (NYR) (Nehru Youth Centre) was developed in 1972.

In 1976-77, in collaboration with the Ministry of Education, the Indian Council of Agricultural Research and the University Grants Commission, a novel scheme called 'Soil Health Care' was started as a part of the NSS programme of the colleges.

Agricultural education in schools

Basic schools

The first bold step to revolutionize the educational system of the country after independence consisted of strengthening basic education. This new system, in addition to its general objectives, aimed at: making education more meaningful by introducing crafts relevant to life and living; focusing on education predominantly based on productivity and work experience; and inculcating in students a sense of, and respect for, the dignity of labour. A national system of education based on this philosophy was expected to bring about renewed confidence and a changed outlook in respect of the reconstruction of society through hard work and self-respect.

The introduction of appropriate basic crafts in the school curriculum and their integration with academic work was the fundamental departure from previous policy for general primary and middle schools. The choice of basic crafts for a particular group of schools was a crucial factor in the scheme. Crafts significant from the point of view of intellectual content, of sequential development of knowledge and skills, and of practical value and efficacy were to be identified and made use of. The curriculum was linked with craft work and with the natural and the social environment. A close integration between the school and the community was envisaged.

Initially, basic education started in the primary schools (classes I-V) but later was extended to the middle level as well (classes VI-VIII). Experience showed that in the first two classes, crafts should not be over-emphasized but rather some light and useful activities, such as clay modelling, paper work, kitchen gardening, etc. should be introduced. The crafts which became popular in the higher classes were spinning and weaving, wood work, gardening and agriculture.

The teacher-student ratio in junior (primary) and senior (middle) schools was about 1 : 30, and normally, there were 1.5 teachers to a class of thirty-five to forty-five students in primary and middle school. When these were converted into junior and senior basic schools respectively, a craft teacher was added in each case. Some additional grants were provided for craft materials, etc. A craft subject was added to the already existing curriculum comprising subjects such as Hindi, mathematics, science and social studies in the junior basic school, and English, mathematics, Hindi, science, social studies and Sanskrit in senior basic schools.

Agriculture as a craft subject was offered in most of the junior and senior basic schools situated in rural areas. Attempts were made to convert the primary and middle schools into junior and senior basic schools as speedily as possible, but in view of the large number of schools involved and the inadequacy of trained teachers in basic education, this could not be achieved quickly. An effort, however, was made to orient the remaining schools on basic lines by introducing socially and culturally useful activities such as social service, cultural and recreational programmes, the Van-Mahotsava,[1] the celebration of national festivals, etc.

Until recently this educational concept was operative in a large number of basic schools. Unfortunately, the ideals of basic education have not been fully realized, since teaching-learning was not craft-centred in the basic schools. In fact, as time went by, the specific craft became one of the normal subjects of the curriculum. This happened because of: (a) an inadequate understanding of the concept of education; (b) the shortage of trained teachers in the philosophy and methodology of basic education; (c) inadequate resources for craft work; and (d) the fact that the high schools (classes VIII/IX–X/XI) could not be redesigned on the basic education model in view of the demands of continuity.

Multi-purpose schools

The multi-lateral or multi-purpose high schools were established, on an experimental basis, in the 1950s in some states following the recommendations of the Secondary Education Commission (1952). The basic aims of the system were: (a) to convert the 'single track' secondary schools/higher secondary schools to multi-purpose schools; and (b) to diversify the educational programmes with a view to meeting pupils' varying needs, interest, talents and aptitudes. The commission, however, did not recommend that all schools should be multi-purpose; there was room for unilateral schools or any variation of such schools as appropriate to the needs of the different states.

Under this system, the approach was to provide more comprehensive courses in both general and vocational subjects. This required the students to select relevant courses in relation to their needs. Nevertheless, there was no intention of giving a particular orientation towards any specialization; rather the students were to receive varied educational experiences around core subjects of general value as well as professional subjects of vocational interest.

The general courses were offered in the first year, and in the second year students were to take optional courses from a number of alternative groups. The curriculum consisted of: (a) two languages, one being the mother-tongue or the regional language; (b) social studies and general science; (c) one craft subject; and (d) three subjects from one of the seven groups, i.e. humanities, science, technical, commerce, agriculture, fine arts, and home sciences. There was sufficient flexibility in the formulation of the curriculum to reflect local conditions. There were two types of options in the agricultural groups—an academic type requiring physics, chemistry, mathematics and biology (college preparatory), and a vocationally biased type requiring practical training in agriculture, applied mathematics and applied sciences (employment preparatory).

This scheme was placed in the state sector and central assistance was made available for strengthening and orienting the high schools/higher secondary schools as multi-purpose schools. Additional funds and trained staff were provided for the new subjects. By 1960/61, 2,115 multi-purpose schools had been started and during the Third Five-year Plan period (1962-66) an effort was made to consolidate

---

1. Tree-planting festival.

the schools rather than to expand them. But this venture did not make much headway because of the following inherent limitations:

At the school stage, the majority of the pupils aspired to proceed to a college education and consequently very few of them opted for vocational streams. Thus, such heavy investment for a small group of students could not be justified.

In most of the schools, no more than three vocational groups could be offered, and hence the envisaged diversification did not take place.

The opportunities provided under the scheme for a wider perspective and orientation were unwittingly utilized for specialization.

The staff, laboratories and facilities could not match the system's aim of generating popular interest.

Successful students found difficulty in finding employment or even gaining admission to higher studies.

10 + 2 + 3 pattern

Although the methodology and approach envisaged in basic education could not be harnessed to the extent expected, its operative principles did influence the educational system. The legacy of basic education, in fact, is now being strengthened by the 10 + 2 + 3 pattern of education (ten years of primary and secondary, two years of higher secondary and three years of college) as recommended by the Education Commission (1964–66). The basic aim is to introduce, together with general education, work experience/socially useful productive work (SUPW) in the school system, starting at primary and proceeding to higher secondary level. At the higher secondary level (+ 2 stage), there are two main streams —academic and vocational. The vocational courses are designed to be terminal in character. However, the scheme is relatively flexible and provides for higher education in the vocational areas after certain remedial courses in basic sciences. Students of the academic stream have the option of taking any course in the vocational stream, and vice-versa. A third stream—combination of academic and vocational—has also been recommended by the National Review Committee on Higher Secondary Education with special reference to vocationalization (1978).

In the area of socially useful productive work and the vocational courses, agriculture and allied fields are expected to find a priority place. Besides the general educational objectives, the new system aims at:

Developing in students the desire to participate in productive activities related to community life and to serve the community.

Imparting skills for the planning and execution of socially useful productive work with a view to making education work-based.

Helping students to explore the world of work and understand the realities of life in order to prepare for a confident entry in to the world outside school.

Studying technologies and related sciences and acquiring practical skills and aptitudes relating to vocations for their economic improvements.

The 10 + 2 + 3 pattern of education has now been accepted as national policy. At the national level, the National Council of Educational Research and Training (NCERT) has played the leading role in providing the necessary guidelines, model syllabuses, textbooks and orientation for the educational planners, administrators, educators and teachers. The Central Board of Secondary Education for the central schools and the state secondary school examination boards for the schools in the states are responsible for implementing the scheme.

Introduction of this new system demands heavy investment in order to develop adequate educational resources, both material and non-material. In view of scarce resources, existing facilities, both material and in terms of staff, are being reoriented in some states to meet immediate needs. Teachers are being trained through in-service courses, summer institutes, seminars and workshops. Initially, the work experience and the vocational courses in some states are being organized in those subjects for which the teachers and other educational facilities are readily available. Additional funds are being provided to supplement existing facilities. In other states, a core teacher in the respective vocational area is being appointed, supported by part-time teachers as and when required.

TABLE 3. School curriculum and allocation of time

Subjects	Time allocation (percentage)		
	Classes I-IV/V	Classes V/VI/VII/ VIII	Classes VII/VIII-IX/ X
Language	20	22	25
Mathematics	20	12.50	12.50
Environmental studies	20	—	—
History, civics and geography	—	12.50	10
Science—an integrated course	—	12.50	16
The arts (music, dancing and painting)	—	10	6 [1]
Socially useful productive work	20	18	18
Games and creative activities	20	12.50	12.50
	100	100.00	100.00

1. Instead of arts, one of the following subjects may be taken: home science, agriculture, commerce, economics, social reconstruction, classical languages, etc.
Source: Report of the Review Committee on the Curriculum for the Ten-year School, New Delhi, Ministry of Education and Social Welfare, Government of India, 1977.

It will be seen from Table 3 that the socially useful productive work (SUPW) occupies almost one-fifth of the total time in all three school systems. SUPW is mainly oriented to agriculture and allied fields, particularly in schools in rural areas. In classes VII/VIII to IX/X, students can choose, in addition to SUPW, vocational subjects such as agriculture, home science, social reconstruction, etc. in place of arts (music, dancing and painting).

At the + 2 stage, students in academic streams devote 15 per cent of their time to language(s), 15 per cent to SUPW and 70 per cent to elective subjects —mathematics, economics, chemistry, etc. In vocational streams, the time spent on language(s) is 15 per cent, on general foundation courses (Gandhian concept of education, agriculture in the national economy, entrepreneurship, marketing, environment protection and development, etc.), 15 per cent, and on elective subjects, i.e. agriculture, business and office management, educational services, home science, etc., 70 per cent.

During the Fifth Five-year Plan (1974-79), the new concept was thoroughly studied and implementation was started. Karnataka became the first state to adopt and implement the system in 1977, while a number of states are still engaged in pre-adoption vocational surveys. Several states introduced this new pattern in the year 1978/79. At present in Karnataka 3,500 students are offered eighty-three vocational courses.

Fishery schools

The importance of fishery education was recognized in the early part of this century in Kerala state. Fishery as an elementary course was introduced in some lower and upper primary schools, and vocational courses in fishery were introduced in three regional technical high schools. These courses were initiated to develop vocational skills in young people in order to support the fishery industries in the state and to prepare them for higher education in fishery, if they so desired.

The courses are offered in the three high school classes VIII, IX and X. The course in fishery is in lieu of the second paper in the regional language. The schools are directly under the Department of Fisheries, government of Kerala, though staff members are drawn from the Education Department.

In class VIII, students are taught introductory fishery, fishery biology—both marine and inland—processing technology, fishery economics, navigation and seamanship and practical work. In class IX, the courses on fishery biology, economics, navigation and seamanship continue. In class X, in addition to fishery biology, navigation and seamanship, fishery economics, courses on pollution of inland waters, craft and gear technology, and processing technology are also added.

These schools have produced a large number of technicians for the fishery industries.

Non-degree programmes

Non-degree educational programmes in agriculture and allied fields have not received the attention they deserve. At the present stage of agricultural growth in India, the ratio between numbers of scientists and technicians is not even 1 : 3, whereas in Taiwan, for example, it is 1 : 7, and in Japan 1 : 10. About 25 per cent of the secondary/higher secondary students proceed to degree-level education; the majority of those remaining must be given an opportunity for suitable certificate/diploma-level training, in order to increase the number of skilled workers and technicians. The broad objectives of non-degree programmes are: (a) to prepare an adequate number of skilled workers and technicians to support scientists in the field and in middle management levels; (b) to provide the necessary knowledge and skills to young people for gainful employment; and (c) to strengthen the foundation for modernizing agriculture and related fields.

Some efforts to organize certificate/diploma courses in different disciplines in the country are worthy of mention: these are illustrative rather than exhaustive.

Agriculture

The Agricultural School, Manjri (1941), was created in order to impart vocational training to rural people. Until 1946, the school offered a one-year certificate course in agriculture; a two-year certificate course was offered from 1947 onwards. The main objectives of these courses in agriculture are to train young farmers in new technology and improved agricultural practices; to provide vocational know-how for those wishing to take up farming as an occupation; and to train students to become progressive farmers and local agents for the dissemination of agricultural information.

The school is headed by a superintendent who is supported by three to six instructors and an equal number of agricultural assistants/demonstrators. Normally staff members are degree holders in agriculture and allied areas. The school, in addition to the other basic facilities, has a farm of 75 hectares, a dairy unit and a poultry farm.

Earlier, the school admitted students who had completed at least seven years of general education and were in the 15-22 age-group. However, since 1974, the minimum eligibility requirement has been matriculation (SSLC). The school's intake capacity is fifty students per year, most of whom receive stipends during their training.

The Agricultural School, Manjri, was transferred to Mahatma Phule Krishi Vidyapeeth (Agricultural University), Rahuri, by the state government of Maharashtra in 1974. Since then, the treatment of the curriculum has become more field-oriented and pragmatic. Trainees are required to devote over 50 per cent of their time to practical work—bullock byre work, animal management, poultry keeping, farm work, etc. and in the remaining time receive theoretical instructions. Through the 'earn-while-you-learn' scheme, students are able to earn some money in addition to having an opportunity to manage a small farm. Table 4 provides a broad view of the school curriculum.

TABLE 4. Syllabus for the two-year school certificate in agriculture

Subject	First year				Second year			
	Work hours		Marks allotted		Work hours		Marks allotted	
	Theory	Practical	Theory	Practical	Theory	Practical	Theory	Practical
Principles and practices of agriculture	99	109	75	50	105	114	75	50
Crop husbandry	99	107	75	50	105	114	75	50
Horticulture	99	108	75	100	105	114	75	100
Animal husbandry and dairy science	99	108	100	50	105	114	100	50
Co-operation and agricultural marketing	99	—	50	—	105	56	50	50
Village administration and land reforms	99	—	50	—	—	—	—	—
Agricultural extension	—	—	—	—	105	56	100	50
Plot work	—	—	—	200	—	—	—	200
Dairy and poultry work	—	—	—	50	—	—	—	50
TOTAL	594	432	425	500	630	568	475	600

In addition, students are involved in extracurricular and co-curricular activities. The medium of instruction is the local language of the state—Marathi. Annual examinations are held and students are required to achieve marks of at least 40 per cent for successful completion of the course. The Manjri Agricultural School, since its inception, has produced 1,551 certificate holders. They are generally employed as *gram sevaks*, assistant *gram sevaks* and agricultural assistants. About 40 per cent of the students are reported to have taken up farming as their occupation. However, in the majority of the schools, the percentage has been much lower.

Agricultural schools on the pattern of the Manjri school multiplied in the different states, as noted earlier. The basic objective, to equip rural people to take up agriculture as an occupation, has not been completely realized; most of the certificate holders have sought white-collar jobs. Many of the schools, moreover, could not provide adequate practical facilities and thus failed to infuse confidence in the trainees for undertaking farming as a vocation or for seeking self-employment.

Animal science

The Indian dairy diploma (IDD) course has a history of over half a century. The course provides trainees with a general background and practical skills in dairy farm management and dairy plant operations, including business aspects of the dairy industry.

Until 1959 a composite general course of two years' duration covering the main branches of dairying was offered. But in order to allow for greater specialization, the course has now been streamed into IDD dairy technology and IDD dairy husbandry. The former course is offered at five centres (Bombay, Anand, Haringhata, Karnal and Bangalore) and the latter at three centres (Allahabad, Haringhata and Bangalore).

The IDD is a two-year course with two specializations—dairy technology and dairy husbandry. A management committee supervises the courses of study and examinations are conducted by a special board. Candidates with higher secondary or equivalent education in science subjects alone are admitted. The course includes both theoretical and practical training, including six months' training on an approved dairy farm/plant.

Since its inception a large number of IDD holders have been produced to support the dairy industry. At present more than 200 diploma holders are produced each year from the different institutions.

In addition to diploma-level courses, there are a number of certificate training programmes of varying duration. The most popular are the courses for veterinary field assistants, stockmen, and veterinary compounders, each being of one-year duration. These courses are provided in about a dozen different institutions.

Fisheries

Training in fishery science gained momentum during the post-independence period. Training facilities for operatives in artisanal fisheries and technicians for managing shore-based establishments were created in several states or union territories. The main responsibility for training technical personnel was entrusted to the Central Institute of Fisheries Nautical and Engineering Training at Cochin. Certificate/diploma courses, namely fishing-boat building (fifteen months), shore mechanics (twelve months), fishing-gear technicians (nine months), shore mechanics and electronic technicians (nine months) are offered. Many other specialized courses are offered by several institutions or government departments. For instance, courses for refrigeration technicians, processing technicians and purse-seine master-fishermen are organized at the Integrated Fisheries Project (Cochin); courses in refrigeration and air-conditioning mechanics by the government of Orissa; courses for fish-processing technicians by the Fisheries College, Mangalore, etc. Table 5 presents details of some of the important courses being organized in different states and union territories.

Agricultural engineering

The Agricultural Engineering Institute, Raichur, Karnataka, of the University of Agricultural Sciences, Bangalore, was the first institute to offer a three-year diploma

TABLE 5. Training centres and courses for operatives for artisanal fisheries

State	Number of training centres	Course/subject	Duration (months)	Seats available	Number of trained persons
Andamans	1	Modern fisheries training	10	20	20
Andhra Pradesh	3	Methods of marine fishing and handling of mechanized boats	12	75	658
Goa	1	Modernized marine fishing operations	10	30	133
Gujarat	3	Marine fishing	10–12	—	1 521
Karnataka	1	Marine fishing	10	30	1 678
Kerala	2	Fishing, management and operation of mechanized marine fishing	10 / 9. 5	30 / 40	1 678 / 3 345
Lakshadweep	1	Fishing, handling of boats and engines	10	20	191
Madhya Pradesh	1	Fisheries management and supervision	.33	—	257
Maharashtra	13	Modern marine fishing, use and handling of machines and equipement	6	20	2 500
Orissa	9	Marine fisheries, inland fisheries and riverine spawn collection	10	110	218
Punjab	2	Inland fish farming	.33	360	320
Tamil Nadu	6	Fishing methods, navigation, seamanship and management of marine diesel engines	10	50	3 500

Source: Fisheries Education and Training in India, Central Institute of Fisheries, Nautical and Engineering Training, Cochin, India, 1977.

course in agricultural engineering (since 1969). The institute provides for two specializations, i.e. farm machinery, and soil and water management. The courses aim at imparting vocational skills in agricultural engineering to support modern scientific farming systems.

This institute is directly managed by the University of Agricultural Sciences. Its academic programme is supervised by the dean, and also by the director of instruction (basic sciences and humanities), director of instruction (agriculture) and director of instruction (veterinary sciences) in their respective faculties. The college is headed by a principal who is supported by twenty-four teaching staff in different disciplines. The institute possesses good facilities in terms of workshops and equipment, instructional farm, library, residential hostels, etc.

The institute enrols forty students annually. Students are eligible for admission after completion of the SSLC examination. The first-year course is focused on basic sciences and humanities; the second year on agricultural engineering—surveying, drawing, workshop, electricity and electrical appliances with limited training in irrigation, drainage, soil conservation, farm structure and tractor opertations. During the third year, students are given intensive training in either of the two specializations. The major part of the training is practical and skill-oriented. The institute follows the trimester system with internal assessment and course credit patterns. For their diploma students have to obtain 144 credits.

Up to 1976/77, six batches of students had completed their courses, comprising 130 diploma holders. Most of the diploma holders have found suitable employment in government agricultural institutions or private organizations. From 1979 the institute proposed to offer two further specializations, i.e. crop processing and dairy technology, and agro-product technology, in order to meet trained manpower needs in these areas.

Home science

Professional courses

Home science courses in the context of the agricultural education system are intended to meet the educational, training and research needs of rural families. Of the twenty-one agricultural universities in the country, eight have home science colleges as constituent units; others have still to follow suit. Most of them offer one- to two-year certificate/diploma courses in para-home science services which also provide a basis for higher education in home science. As an illustration, details of a two-year diploma course being offered at the Gobind Ballabh Pant University of Agriculture and Technology, Uttar Pradesh, are shown in Table 6.

TABLE 6. Course structure of home science diploma (two years)

Main fields	Number of courses	Credit hours		
		Total	Theory	Practical
Child development and family relationships	9	27	25	2
Clothing textiles	11	34	16	18
Foods and nutrition	12	37	18	19
Home management	13	40	23	17
Extension education	4	9	6	3
Languages	6	12	8	4
Basic sciences	8	27	19	8
Social sciences and humanities	2	5	3	2
Agriculture and animal sciences	3	8	3	5
Work programme	1	3	—	3
TOTAL	69	202	121	81

The diploma course is open to high school girls who may later proceed to the four-year B.Sc. home science degree programme. The main objectives of the course are: (a) creating interest in a home science profession, especially among rural girls/women; (b) providing scientific knowledge and practical skills for effective home management; (c) imparting competence for gainful employment either at home or outside and for fuller utilization of available resources in the welfare of families; and (d) developing in young women a desire and a capacity to enrich their personal and family lives and all aspects of daily life.

The Home Science College is one of the constituent units of the university. The college is headed by the dean of home science, who is supported by adequate scientific and technical staff.

Diploma students are required to complete 107 credit hours of work during six trimesters. This also includes co-curricular and extracurricular activities. The main branches of study and the credit distribution are shown in Table 6.

There is a big demand for para-home science diploma holders and graduates in the fields of rural development, agriculture, social welfare and public health. Workers trained in home science such as *gram sevikas* (VLWs), *bal sevikas*, *anganwadi* workers, health visitors, health extension educators, adult educators, etc. are not readily available.

Courses for housewives

Gujarat Agricultural University has a Home Science School at Anand Campus which offers a special two-year diploma course. The basic aim of this course is to provide knowledge and skills to rural girls with regard to home management and family relationships.

Girls between 16 and 25 years of age with education up to class VIII are eligible for admission to this course. The course curriculum does not include basic sciences, language or social sciences and humanities, since the intention is not to encourage these girls to proceed to higher education. In addition to the home science courses, a number of courses are provided in agriculture and allied areas specially suited for farm women, e.g. kitchen gardening, animal husbandry, dairy science, poultry and horticulture. Since 1960 this school has produced 326 diploma holders. It is reported that girls having this two-year diploma in home science are favoured as wives by graduates and postgraduates.

Bakery courses

There are seventeen institutions which give bakery training. They include the Pilot Bakery Project of the University of Agricultural Sciences, Bangalore, the School of Baking of the Gujarat Agricultural University, Anand, four catering colleges and eleven food craft institutes.

The Pilot Bakery Project (1968) at Bangalore provides training in modern techniques of baking and caters for the training needs of southern states. The project has five scientific staff members and functions as an integral part of the University of Agricultural Sciences. It offers a twenty-week certificate course for practising bakers having the matriculation or its equivalent and for persons having one year of practical experience in bakery. The course of study furnishes a basic knowledge of raw materials and involves simple arithmetic and calculation of ingredients, the physics and chemistry of baking, economics, hygiene, nutrition, sanitation and safety, the bakery oven and its maintenance, operation of semi-automatic plant, and preparation of various bakery products.

Since 1968, nineteen courses have been organized and 221 persons trained. The project also runs special and short training courses and conducts related research studies.

The pilot project operates on the model of the School of Baking of the Gujarat

Agricultural University, which was established as a part of the Institute of Agriculture˙at Anand (1963) in collaboration with Wheat Associates of the United States. The school has already trained 366 persons besides organizing a number of short courses for different groups.

Forestry

Four rangers' colleges have been established to provide technical education and training in forestry: at Dehradun, Uttar Pradesh; Coimbatore, Tamil Nadu; Kurseong, East Bengal; and Chandrapur, Maharashtra. These colleges offer a two-year diploma course designed to prepare staff for planning and supervising general forestry work: logging, timber operation, soil conservation, construction of roads and buildings, maintenance of records and accounts, etc. After reaching the age of 18 and having taken the intermediate or pre-university or higher secondary examination in science, trainees are selected on the basis of their performance in a written test, an interview, an endurance test and a medical examination. Some field experience is also required.

The forest rangers' colleges are under the Central Sector in the Union Ministry of Agriculture and Irrigation. The Director of Forest Education stationed at Dehradun is in overall charge of forestry education and training in the country. The dean is the head of the Rangers College. The other academic staff of the college are lecturers and instructors who normally belong to the All-India Forest Services. Visiting staff on a short-term basis are also used for specific courses.

Training includes both theoretical and practical aspects of forestry science. The broad areas of the curriculum are as follows: forestry and allied subjects, with 315 lectures; forest utilization, 103; forest engineering, 150;[1] physiography and geology, 30;[1] forest botany, 147;[1] procedures and accounts, 40; forest law, 46; wildlife management, 22; and tribal welfare, 10.

Out of the total period, twenty-eight weeks are devoted to theory and practical work, fourteen weeks to educational tours and exercises and the rest to vacation, examinations, etc. In addition, trainees are given physical training, first-aid instruction, and weapon training. The intake capacity in each college is 80–90 per year.

Rural oriented courses

The Shrimali Committee recommended the establishment of special institutes in rural areas for rural higher education. Consequently, the central government started fourteen rural institutes in 1956 in different states. They were associated with institutions which had deeper roots in rural services. The main objects of such an educational venture were: (a) to provide rural oriented higher education to rural youth; (b) to bring higher education closer to the villages in order to attract more rural youth to education; and (c) to develop in rural youngsters a new type of leadership and citizenship.

The original intention was to develop such rural institutes for technician-level education and training, but they are now being upgraded to provide degree

---

1. Practicals in addition.

programmes as well. Diploma courses in various disciplines, based on local needs, have been offered in the rural institutes; e.g. the Gandhigram Rural Institute offers diploma courses in preparatory agriculture, rural services, sanitation, health and hygiene, etc. for matriculated students. Initially these institutes were financed by the central government's Ministry of Education but later became the responsibility of the states. Local management of the institute rests with the Governing Body/Management Committee. The training programmes are practical and vocationally biased with due support of basic sciences, social studies and the humanities.

Higher agricultural education

Prior to the creation of the agricultural universities, higher education in agriculture and allied fields was undertaken by colleges affiliated to the general universities. Such colleges functioned mainly for instructional purposes. The Indian Agricultural Reserach Institute—deemed to be a university—first adopted the major principles of the agricultural university concept in its academic programme for postgraduate degrees in 1958. Following the recommendations of the University Education Commission and the two joint Indo-American Teams, the creation of agricultural universities became national policy—a decision of far-reaching importance and implications. At present there are twenty-one agricultural universities located in sixteen major states of the country (see Fig. 3).

In an agricultural university, teaching, research and extension are fully integrated. The main objectives of such universities are: (a) imparting higher education in agriculture and allied sciences; (b) encouraging scientific progress through research studies; (c) organizing extension education in selected communities/districts to serve as a model for the respective state extension machinery; and (d) providing overall leadership in the matter of agricultural development, including agricultural education at the lower levels.

The Indian agricultural universities are autonomous institutions that are mainly guided and supported by the Indian Council of Agricultural Research and the state governments. By and large, these universities are patterned on the Model Act of ICAR (1966). Of the twenty-one agricultural universities, six are mono-campus type and fifteen have multi-campuses, the largest being Jawaharlal Nehru Krishi Vishwa Vidyalaya, Madhya Pradesh. In addition, there are thirty-four agricultural colleges affiliated to the general universities, including two private agricultural colleges in Maharashtra state which are affiliated to Punjabrao Krishi Vidyapeeth, Akola (agricultural university). Each agricultural university normally consists of constituent colleges in agriculture, animal sciences, including veterinary science, basic sciences, agricultural engineering, and home science. Recently some universities have started offering special courses/degree programmes in fisheries, horticulture, food technology, water technology, forestry, agricultural meteorology and business management.

All the universities have adopted the course credit, internal assessment, and semester/trimester systems. Generally, the authorities of the universities are a board of management, an academic council and a board of studies. The Research Council and the Extension Education Council have also been incorporated into the university system. As an example the broad structure of Punjab Agricultural University (PAU) is shown in Figure 4.

Elements of the structure and terminology of agricultural education in India

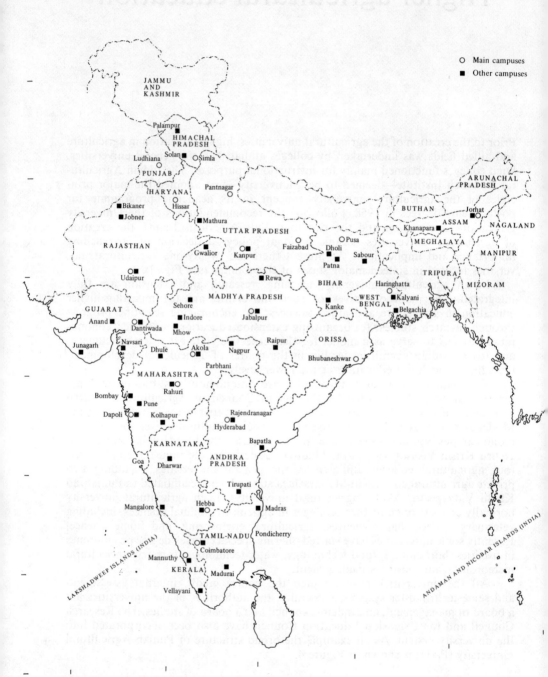

FIG. 3. Agricultural universities in India.

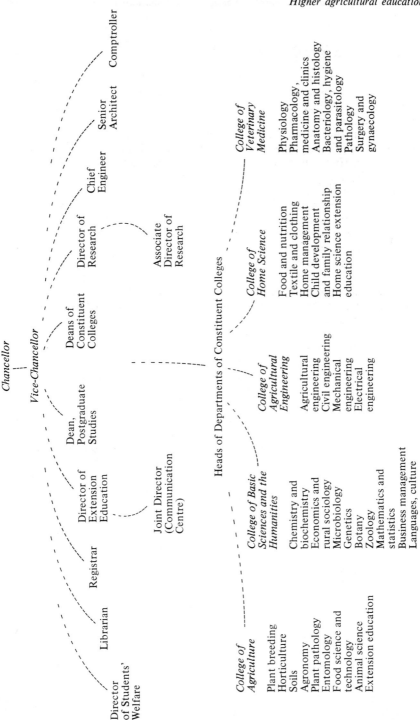

FIG. 4. Organizational structure of the Punjab Agricultural University.

Graduate programmes

The graduate programmes in agriculture and allied areas came into being at the beginning of the twentieth century. The degree programme in agriculture started in 1909 in the College of Agriculture, Poona; animal sciences/veterinary science were introduced in 1936 at Madras Veterinary College, Madras; agricultural engineering in 1942 in Allahabad Agricultural Institute, Allahabad; home science (in the context of the agricultural university) in 1963 in Andhra Pradesh Agricultural University, Hyderabad; and fisheries in the University of Agricultural Sciences, Bangalore in 1969. Recently, B.Sc. (agriculture honours), B.Sc. (horticulture) and B.Sc. (dairy technology) have also been introduced in a few institutions. The objects of graduate education in agriculture and allied areas are preparing potential graduates to: (a) take up employment in higher education; (b) join government or non-government organizations that promote agricultural development; and (c) return to their farms to take up agriculture as an occupation and play a leadership role in the dissemination of the latest technical know-how.

The duration of the degree programmes varies from one situation to another. For instance, in southern India, after the school stage (eleven/twelve years), four years are required for a degree in agriculture and five years for a degree in agricultural engineering and veterinary/animal sciences. In the north, it usually takes three years for a degree in agriculture after achieving intermediate level in science or in agriculture. Nevertheless, the latest trend is to provide a four-year degree programme in agriculture following pre-university training. In animal science and agricultural engineering, the duration is four years after achieving intermediate level in science. The fishery degree programme requires four years after the pre-university science course; and the home science degree three years. Where an honours programme is available (PAU), this requires a further year of study following the undergraduate course.

Admissions are mainly on a merit basis (minimum 50-60 per cent marks) with some relaxation for weaker sections of society. In Rajendra Agricultural University, Pusa, Bihar, a competitive examination is held for admission to graduate programmes in agriculture and animal sciences.

Though English is the medium of instruction in most cases, a few universities also use regional languages. The average percentage of credit hours in basic science and humanities, core professional courses and electives, as also theory and practicals in different degree programmes, is shown in Table 7.

TABLE 7. Average percentage of credit hours offered in different degree courses at various agricultural universities during 1975/76

Degree programmes	Basic science and humanities	Core professional courses	Electives	Theory	Practical
B.Sc. (agriculture)	29	67	4	63	37
B.V.Sc.	5	93	2	68	32
B.Sc. (agricultural engineering)	24	70	6	66	34
B.Sc. (home science)	28	68	4	65	35
B.Sc. (horticulture)	19	75	6	61	39
B.Sc. (fisheries)	21	68	11	57	43
B.Sc. (dairy technology)	24	76	—	62	38

Source: Report of the Review Committee on Agricultural Universities, New Delhi, ICAR, 1978.

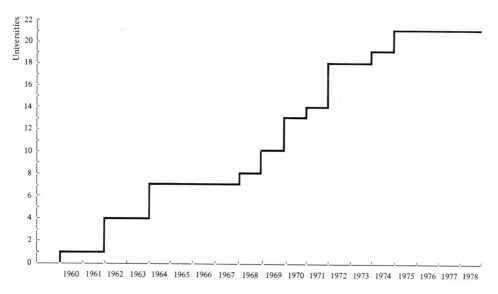

FIG. 5. Growth of agricultural universities in India.

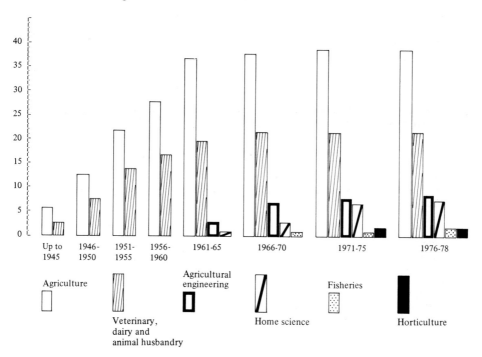

FIG. 6. Growth of colleges in agricultural universities in the fields of agriculture and allied areas.

There are some local variations in the curricula structure of agricultural universities in their fields of specialization. As illustration, the curricular structures of a few major degree programmes are given in Tables 8 to 12.

TABLE 8. Distribution of credit hours in the major subjects of the four-year degree programme in agriculture (PAU)

Basic areas	Number of courses	Credit hours		Total credits
		Theory	Practical	
Basic sciences	13	32	12	44
Agronomy	5	13	6	19
Soil science and biochemistry	4	12	4	16
Horticulture	3	8	3	11
Agricultural engineering	1	3	1	4
Agricultural economics	3	8	1	9
Livestock production and management	2	4	2	6
Microbiology	1	3	1	4
Plant breeding and genetics	3	8	3	11
Plant pathology	1	4	1	5
Entomology	1	3	2	5
Food technology	2	4	2	6
Extension education and rural sociology	3	7	1	8
Forestry and ecology	2	4	1	5
Statistics	1	3	—	3
Practical field training	12	—	24	24
Electives and allied courses	—	—	—	44
TOTAL	57	116	64	224

TABLE 9. Distribution of credit hours in major subjects of the four-year degree programme in veterinary and animal husbandry (PAU)

Basic areas	Number of courses	Credit hours		Total credits
		Theory	Practical	
Veterinary anatomy and histology	6	13	10	23
Livestock and poultry management, animal breeding and nutrition (animal science)	9	24	9	33
Food technology	2	5	2	7
Veterinary physiology	6	18	6	24
Veterinary bacteriology and hygiene	5	16	5	21
Veterinary pathology	5	11	5	16
Veterinary gynaecology and obstetrics	4	8	4	12
Veterinary medicine	7	19	4	23
Veterinary survey and radiology	5	11	7	18
Veterinary clinical practice	8	3	12	15
Veterinary pharmacology	3	11	3	14
Veterinary parasitology	3	9	3	12
Extension education	2	5	2	7
Agronomy	1	2	1	3
Economics of livestock management	1	2	1	3
Fundamental of genetics	1	3	1	4
Principles of statistics	1	3	—	3
TOTAL	69	163	75	238

TABLE 10. Distribution of credit hours in the major subjects of the four-year degree programme in agricultural engineering

Basic areas	Number of courses	Credit hours		Total credits
		Theory	Practical	
Basic sciences	9	30	7	37
Agricultural engineering	15	28	17	45
Civil engineering	9	23	9	32
Mechanical engineering	12	26	13	39
Electrical engineering	4	14	4	18
Surveying	1	—	1	1
Soils	2	6	2	8
Agronomy	1	4	1	5
Animal science	1	4	1	5
Entomology	1	3	1	4
English writing	2	2	3	5
Economics	2	6	1	7
Microbiology	1	2	1	3
Extension education including rural sociology	2	5	1	6
Statistics	1	3	—	3
History	1	3	—	3
Sport and physical training	7	—	7	7
Workshop	1	—	1	1
TOTAL	72	159	70	229

TABLE 11. Distribution of credit hours in the major subjects of the four-year degree programme in home science

Basic areas	Number of courses	Credit hours		Total credits
		Theory	Practical	
Basic sciences	11	22	11	33
Agriculture (horticulture and entomology)	4	8	4	12
Animal science	2	4	2	6
Behavioural sciences including economics	5	16	—	16
Language and culture	3	8	—	8
Applied and fine arts	5	5	7	12
Child development	6	17	3	20
Clothing and textiles	6	8	10	18
Foods and nutrition	5	9	9	18
Home management	5	11	7	18
Home science education and extension	5	10	3	13
TOTAL	57	118	56	174

TABLE 12. Distribution of credit hours in major subjects of the four-year degree programme in fisheries

Basic areas	Number of courses	Credit hours		Total credits
		Theory	Practical	
Basic sciences and humanities	17	23	18	41
Fishery biology	11	22	15	37
Aquaculture	4	8	7	15
Fishery hydrography	88	18	7	25
Fishery economics	2	5	2	7
Fishery statistics	2	3	1	4
Fishery processing technology	8	12	12	24
Fishery microbiology	2	4	3	7
Fishery biochemistry	1	2	2	4
Fishery engineering	6	10	7	17
Fishery extension	1	2	1	3
Others: (major field of specialization (4) electives (2) and study tour (2))	3	—	—	8
TOTAL	65	109	75	192

Postgraduate education

Generally speaking, postgraduate programmes in agriculture and animal sciences started with diploma courses. Postgraduate degree programmes, in most cases, were only instituted after 1930. The main purposes of the postgraduate programmes are: (a) to train specialized scientists in different disciplines for supporting teaching, research and extension programmes of agricultural and allied institutions/agencies; (b) to increase knowledge through postgraduate studies and investigations; and (c) to provide professional leadership in the discovery and dissemination of scientific innovations.

In almost all agricultural universities/colleges, masters degree programmes in agriculture, horticulture and veterinary sciences are offered. Ph.D. programmes have also been introduced in these disciplines but in fewer institutions. In other disciplines postgraduate degree programmes are very limited. An overall view of the postgraduate programmes offered is given in Table 13.

The dean of postgraduate studies is the executive head of the academic programme. In some universities, the dean of the relevant faculty looks after both graduate and postgraduate instruction.

Normally only first or high second-class B.Sc. graduates in the respective disciplines are admitted to postgraduate degree programmes. Each university has its own procedure for screening and selecting students. Tamil Nadu Agricultural University (TNAU), Coimbatore, for instance, constitutes a committee headed by the dean of the faculty to assess the suitability of the candidates, with 50 per cent weightage given to the candidate's academic performance in his B.Sc. degree and the rest to service, publications, quiz and oral test.

TABLE 13. Number of institutions offering postgraduate programmes

Faculty	Number of universities/ institutions			Number of colleges
	B.Sc.	M.Sc.	Ph.D.	
Agricultural	21	21	16	38
Veterinary	18	19	15	21
Agricultural engineering	9	4	3	9
Home science	8	5	1	8
Fisheries	2	2	1	2
Horticulture	3	21	14	3
Dairy science	3	2	3	1
Basic science	1	4	2	7
Business management	—	1	1	—
Agricultural statistics	—	1	1	—

Source: Report of the Review Committee on Agricultural Universities, op. cit.

An advisory committee is created for each student to supervise his academic work. In addition to course work, the students must undertake research projects for their theses. Students are required to complete successfully forty-five to sixty credits of course work besides their theses for the M.Sc. or Ph.D. degree. The courses are divided into major and minor fields of specialization with a few compulsory courses, e.g. statistics. In general, examinations are conducted internally but for Ph.D. degrees thesis work and the oral examination are assessed by external examiners. Students are required to conduct two to four seminars in their major and minor fields of study as well as on their research work.

Enrolment and employment

The number of educational institutions and their various departments has steadily grown. The production of agricultural and allied graduates and postgraduates has increased in the same propostion. Table 14 shows 1975/76 admission capacity and output of graduate and postgraduate students in defferent faculties.

The curricula in agriculture and allied fields have been criticized as being too academic and too theoretically biased. Self-employment in these fields has, therefore, been limited. According to the 1971 census report, only 4 per cent of agricultural graduates were self-employed: of the rest 12 per cent were employed in the private sector and 84 per cent in the public sector. It has been reported that, in 1973, over 10,000 unemployed agricultural graduates and postgraduates were registered and looking for work. During the same period, a surplus of 1,915 veterinary graduates and 600 postgraduates was reported. The output of graduates in other disciplines has been relatively limited and thus employment opportunities have been more satisfactory. A concerted effort is now being made to introduce production-oriented courses and 'earn-while-you-learn' projects in the graduate programmes of the agricultural universities in order to make the graduates more employable.

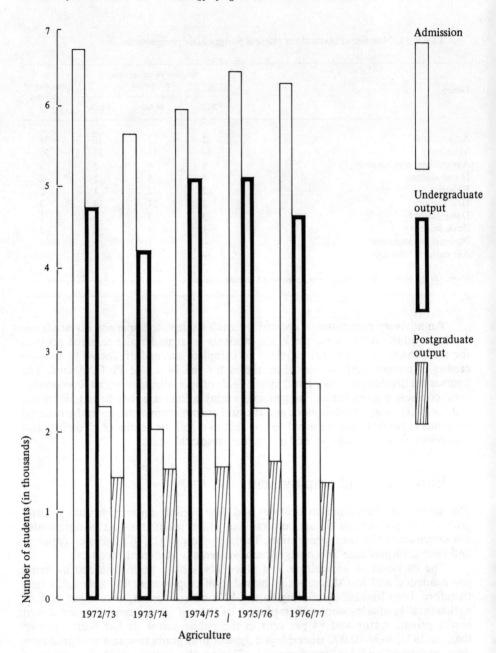

FIG. 7. Admission and output in undergraduate and postgraduate agriculture and allied education.

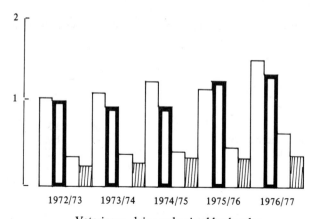

Veterinary, dairy and animal husbandry

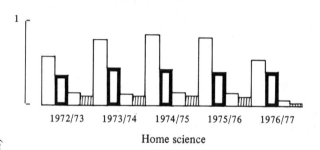

Home science

Number of students (in thousands)

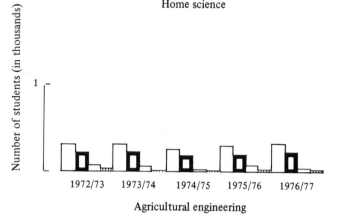

Agricultural engineering

TABLE 14. Total overall admission and output in B.Sc. and M.Sc. courses in different agricultural university faculties, 1975/76

Subject number	Faculty	Admissions		Output	
		B.Sc.	M.Sc.	B.Sc.	M.Sc.
1	Agriculture	4 114	1 665	3 314	1 091
2	Veterinary and animal husbandry	1 141	414	1 053	176
3	Agricultural engineering	395	28	288	5
4	Home science	528	53	184	16
5	Horticulture	80	—	60	—
6	Fisheries	40	16	18	—
7	Dairy technology	60	—	39	—
8	Basic science and humanities (PAU only)	18	40 [1]	40	25
9	Other courses	—	23 [2]	—	17 [3]
	TOTAL	6 376	2 221	4 996	1 330

1. Approximate figures.
2. A total of 18 students admitted in business administration at Punjab Agricultural University (PAU) and 5 students admitted in environmental biology at Tamil Nadu Agricultural University (TNAU).
3. Output in business administration at PAU.
Source: Report of the Review Committee on Agricultural Universities, op. cit.

A study carried out by the Department of Agriculture of the Ministry of Agriculture and Irrigation, Government of India, estimated that there were 72,500 agricultural graduates including about 18,000 postgraduates by the end of 1973. Table 15 shows the employment pattern of agricultural graduates.

TABLE 15. Employment pattern of agricultural graduates and postgraduates

Subject number	Activity	Number employed	Percentage
1	*State government*		
	Block headquarters	5 700	8.0
	General administration	33 500	46.5
2	Central government departments	500	0.7
3	Agricultural universities and colleges	5 000	7.0
4	Research stations (ICAR)	5 000	7.0
5	Agricultural schools including veterinary schools	3 000	4.2
6	Central establishment and corporation	1 900	2.5
7	Commercial banks	2 400	3.1
8	*Postgraduate education*	4 000	5.6
		61 000	84.6
	Unaccounted balance	11 000	15.4
	TOTAL	72 000	100.0

Source: Report of the Review Committee on Agricultural Universities, op. cit.

A recent study (1977) carried out by the Institute of Applied Manpower Research on 'Agricultural Manpower, Planning and Utilization', has estimated the demand for agricultural graduates during the next twenty years. These estimates are shown in Table 16.

TABLE 16. Estimated demand for agricultural graduates for coming two decades

Subject number	Area of activity	Employment size		
		Current	Optimum	Increase
1	Block	6 000	40 000	34 000
2	State and central administration	34 000	68 000	34 000
3	Regional research stations of the universities	—	5 000	5 000
4	Animal husbandry function	—	15 000	15 000
5	District-level subject-matter specialists	—	3 000	3 000
6	Teachers in secondary schools	2 300	35 000	32 700
7	Research at central stations	5 000	—	—
8	Commercial banks	2 400	24 000	21 600
9	Teachers in agricultural colleges and universities	5 000	10 000	5 000
10	Central establishment and corporation	2 000	5 000	3 000
	TOTAL			153 300
	Attrition requirement			28 000
	Total supply necessary			181 300

Source: Report of the Review Committee on Agricultural Universities, op. cit.

Postgraduate diploma courses

Postgraduate diplomas generally preceded the introduction of postgraduate degree programmes in several institutions/colleges. In the absence of adequate specialized training facilities at postgraduate level, postgraduate diplomas in different disciplines were intended to meet the needs for trained manpower in agricultural institutions and industries. The two-year diploma course leading to an associateship in agriculture was first started in 1923; it was reorganized and strengthened in 1945 and continued until 1958 at the Indian Agricultural Research Institute. Some of the popular postgraduate diplomas now being offered are as follows:

Diploma in dairy engineering

This course imparts practical training in the maintenance of dairy plants and provides a theoretical background in general dairying and dairy plant machinery. It is a one-year course and admission is open to candidates who have completed a degree/diploma course in electrical/mechanical/chemical/agricultural engineering.

The course is offered at the National Dairy Research Institute (NDRI), Karnal, with five to ten admissions every year.

Diploma in sheep and wool husbandry

The duration of this course is nine months only. The course is designed to give a comprehensive practical background in sheep and wool production so that trained persons can be profitably employed in the appropriate departments and institutions. There are only ten places for candidates having gratuated in animal/ agriculture/veterinary science. The training programme includes almost all aspects of sheep and wool production—Indian sheep and wool industries and practices, improvement through better management, fodder and forage productions, scientific feeding and breeding, health cover and evaluation of the produce.

The course is offered at the Central Sheep and Wool Research Institute, Avikanagar, Rajasthan.

Diplomas in agricultural statistics

Three postgraduate diploma courses, each of one-year duration, namely, the senior certificate course, the professional statistician certificate course and the diploma course are offered at the Indian Agricultural Statistics Research Institute, New Delhi. The eligibility requirements for these course are a B.Sc. degree in agriculture or an allied field, M.A./M.Sc. degree in statistics or mathematics and professional statistician certificate course, or M.Sc., in agricultural statistics respectively.

The training programme of these courses consists of lectures and laboratory practicals (computational exercises). In the senior certificate course, the topics covered are descriptive statistics, statistical methods, design and analysis of experiments, sampling techniques, agriculture and animal husbandry statistics, statistical genetics and bio-assays, demography, statistical quality control, and economic statistics and data processing and computer programming. In the professional statistician certificate course, in addition to the above subjects, advanced statistical methods and econometrics are also included.

For the diploma course students are required to undertake a research project and submit a thesis at the end of the training programme.

Diploma in sericulture

The Central Sericulture Research and Training Institute (Mysore), Government of India, offers a course in the science and practice of sericulture leading to the award of the diploma in sericulture. This is a fifteen-month course and is open to graduates of recognized institutions with biology (botany or zoology) or agriculture as one of the major subjects. The Academic Council, headed by the director of the institute, is responsible for the conduct of this course.

The admission capacity is only twenty-five per annum and the training programmes are divided into two sessions, including three months' village training. The medium of instruction is English. Students are required to complete courses for 1,900 marks (1,000 theory + 900 practicals) in silkworm rearing, biology of mulberry plants, soil science, silk recting, economics and statistics, and organization of the silk industry.

One-year diploma in business management

Recognizing the scope and importance of managerial requirements in agricultural business, the Tamil Nadu Agricultural University (TNAU) initiated a postgraduate diploma course in business management in 1976 in its faculty of agriculture. Candidates desirous of admission must possess B.Sc./M.Sc. in agriculture and allied areas or economics, commerce and business management with three years of experience in agricultural finance, agricultural marketing, agro-industries or agricultural extension service in every case.

Students are required to complete forty-eight credit hours during the academic year—thirty-six credits for courses and twelve credits for project work.

Two-year diploma in business management

The Indian Institute of Management, Ahmedabad, established in 1962, initially offered a one-year residential programme for management (PMA) from 1970. One year proved inadequate and a two-year integrated postgraduate diploma programme for management in agriculture was therefore initiated in 1974/75. The main objective of this programme is to develop professionals in management to cater for the requirements of agricultural departments and institutions.

The course is open to candidates with a bachelor's degree or its equivalent in agriculture or allied subjects. The first-year curriculum is designed to provide basic knowledge, understanding and skills that are deemed essential for professional managers. The first-year courses comprising business policy, economics, finance and accounting, marketing, organizational behaviour and production and quantitative methods are, therefore, compulsory for all students. All second-year courses are electives except for business policy (classes I and II). Students specializing in agricultural management have to take additional courses such as agricultural marketing, environment, agricultural finance, agricultural development policy, computer and data processing systems, organizational design, etc. Elective courses are also available in the agricultural sector.

Diplomas in fisheries

In the absence of higher education in fisheries, postgraduate diplomas/certificates were available for technical, managerial and scientific positions. Three postgraduate diplomas, in fisheries science, in fish processing technology and in fish culture technology, have been organized at the Central Institute of Fisheries Education, Bombay (first course), and Fisheries College, Mangalore (latter two courses) respectively. The first course is of two years' duration, whereas the latter two are each of one year. Two postgraduate certificate courses—in inland fisheries development and administration, and in fisheries extension—are offered at the Inland Fisheries Training Unit, Barrackpore and the Central Fisheries Extension Training Centre, Hyderabad, respectively. The duration of the first course is one year and of the second only ten months.

The entry qualification for all these courses is graduation in science, but for the diploma in fish-culture technology the requirement is a B.F.Sc. The medium of instruction is English in all cases and the intake capacity varies from twenty-five to sixty per annum.

Diploma in forestry

A diploma in forestry has been offered to Indian forest service probationers since

1926 at the Indian Forest College, Dehradun. The probationers are recruited through an all-India competition from among science graduates. The course is of two years' duration and, on completion, students are awarded the diploma of the college and are known as Associates of Indian Forest College (AIFC). Before joining the service, they are given four months' further training in administration and management.

Probationers are given both theoretical and practical training in the proportion of 2 : 1. In addition, educational tours and excursions are organized regularly, occupying about 40 per cent of the total time.

Diploma in social service with specialization in integral rural development

The Xavier Institute of Social Service, Ranchi, Bihar, has recently initiated a post-graduate diploma course in social service with specialization in integral rural development. This course prepares students for employment as community development organizers, field researchers or executives for government and non-government organizations.

The eligibility requirement for admission is a degree or its equivalent in arts, science or commerce. The course is of two years' duration and comprises such varied subjects as behavioural science, agriculture, research methodology, rural extension, rural banking, rural development, etc. Fifty students are admitted annually. Government policy for increased attention to integrated rural development has made the course increasingly popular.

Teacher training in agriculture

The provision of teacher education in agriculture has been limited. Many ad hoc teacher training programmes, e.g. short training courses, workshops and seminars, have been organized by teacher training colleges, agricultural colleges and universities. Programmes of teacher training cover both technology and pedagogy. Some institutionalized approaches to teacher training in agriculture are briefly discussed below.

Regional college of education

The four regional colleges of education were established following the introduction of diversified curricula in multi-purpose higher secondary schools. These colleges are run by the National Council of Educational Research and Training (NCERT) —an autonomous body under the Union Ministry of Education, Social Welfare and Culture, Government of India. The principal is the executive head and there are eight to ten departments including a full-fledged department of agriculture.

These colleges have been organizing varied programmes in agriculture with a view to improving the ability of teachers to teach agriculture in the school system. They offer in-service training programmes of short duration in the latest agricultural technologies and methods of teaching. Seminars and workshops for developing agricultural teaching materials are also organized. In the pre-service training programme, integrated four-year science courses leading to B.Sc. and B.Ed. degrees include agriculture in the scheme of work experience. In the one-year B.Ed. degree programme, agriculture is included as one of the vocational subjects. In the case of the B.A.Ed. programme at the Regional College, Mysore, work experience in horticulture is given in the first, third and fourth years of the academic programme.

A one-year B.Ed. course exclusively for agricultural teachers was formerly offered in these colleges, but this was discontinued in view of the few agricultural teachers required by the schools.

Extension education institute

The extension education institutes, three in number, came into being in order to provide in-service training for the faculty of *gram sevak/gram sevika* training centres and later the farmers' training centres. The training programmes of the institutes were mainly focused on training in curriculum development, teaching methodology, preparation and use of audio-visual aids and mass communication.

These institutes are directly controlled and financed by the Directorate of Extension, Ministry of Agriculture and Irrigation, Government of India, in close association with the agricultural universities/institutes in the states in which they are located. Each institute is headed by a principal and has faculty members in the fields of extension education, agricultural engineering, educational psychology, audio-visual aids, animal husbandry, agricultural economics and home science.

Usually two types of course, regular and ad hoc, are offered by the institute. The regular courses vary from two weeks' to three months' duration. The training courses organized at the extension education institutes up to 1975 are shown in Table 17.

Teacher training centres

The teacher training centres established in 1976 by ICAR represent a major landmark in the training of non-degree level teachers in agriculture and allied fields. The basic object is to develop rich practical facilities in specialized areas for imparting training to teachers based mainly on work experience, on the principle of 'learning by doing' and 'teaching by doing'. Participants in the teacher training programmes are mainly teachers of the Krishi Vigyan Kendras (KVK), [1] teachers/trainers from the farmer training centres, teachers in the extension education institutes, and schoolteachers dealing with work experience in the secondary schools and vocational courses in higher secondary schools.

At present there are seven teacher training centres in such specialized fields as those referred to above; a few more such centres are expected to be established during the Sixth Five-year Plan (1978–83). These training centres are associated with the specialized research institutes of the Indian Council of Agricultural Research in order to ensure that they have strong scientific back-stopping. The centres work under the overall administrative structure of the respective institutes and are headed by the senior scientists/chief training organized. The other staff consists of the training organizers, training associates and training assistants in the specialized fields. Overall supervision and monitoring are the responsibility of the Agricultural Education Division of ICAR.

The salient features of the academic programmes of the centres are: (a) rich subject-matter content (almost 80 per cent); (b) integrated pedagogical content (20 per cent) and work experience as a method of training (almost 75 per cent) and lectures/discussions (about 25 per cent). The duration of the courses varies with the specific needs of the particular group of teachers.

In view of the innovative approach being followed in these teacher training centres, it may be helpful to cite an example of the organization and conduct of a training course. At the teacher training center, Kasturbagram, an agricultural orientation course for home-science teachers of KVKs was organized for thirty-six days. Such an orientation course for women teachers of KVKs was considered essential in view of their training responsibility for farm women and the transfer of technologies in the rural areas. The curriculum included relevant content of crop and livestock production, pedagogy, and the concept of KVK training. The subject weightage and the ratio between theory and practical may be seen in Table 18.

1.   Institutions set up by ICAR in the agricultural universities to promote research in agronomy.

TABLE 17

| Course | Duration | Extension education institutes | | | | | | | |
| | | Nilokheri | | Anand | | Rajendranagar | | Total | |
		Number of courses	Persons trained	Number of courses	Persons trained	Number of courses	Persons trained	Number of courses	Persons trained
Training course in extension teaching methods, techniques and communication media for the instructional staff of GTCs/HSWs agricultural schools	3 months 2 months 1.5 months	32	622	24	402	22	344	78	1 368
Refresher course/workshop in extension teaching methods, etc. for the instructional staff of GTCs/HSWs agricultural schools	2 weeks	5	74	4	46	2	36	11	156
Refresher course for the instructional staff of FTCs (men and women)	2 weeks	7	95	8	138	8	138	23	371
Workshop in education administration and teaching methods for principals of GTCs and agricultural schools, etc.	2 weeks	8	125	—	—	—	—	8	125
Specialized training course for demonstrators of GTCs/HSWs and agricultural schools	4 weeks	1	20	—	—	—	—	1	20
Training course in audio-visual aids and equipment for publicity assistants and demonstrators of GTCs, etc.	2-3 weeks	1	10	1	13	1	17	3	40

Source: Extension Training for Rural Development, New Delhi, Farm Information Unit, Directorate of Extension, Ministry of Agriculture and Irrigation.

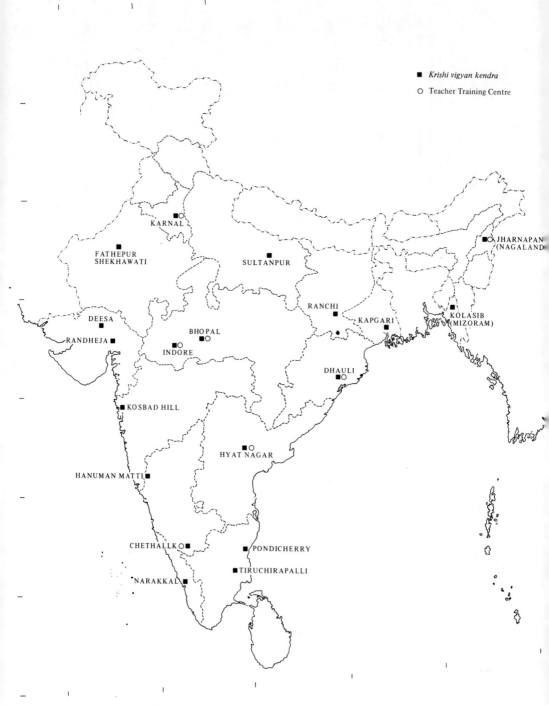

FIG. 8. Krishi Vigyan Kendras and teacher training centres established during 1976/77.

TABLE 18. Subject-wise weightage and ratio between theory and practical on the basis of number of hours involved

Subject number	Subject	Number of hours		Total
		Theory	Practical	
1	Crop-production technology	12	25	37
2	Live-stock production	7	13	20
3	Rural development including village surveys and crafts	6	32	38
4	Pedagogy including development and demonstration of lesson plans, library work, etc.	6	51	57
5	General orientation about KVK pattern of training and participation in community activities such as weekly community cleaning and daily prayer. *Sangeet* (music) for cultural and other activities	5	21	26
		36	142	178

The ratio of theory to practical works out at 1 : 4 and the weightage given to pedagogy, including library work, in relation to technology was 31 per cent. Both teachers and trainees made an evaluation of the course and identified areas for improvements in future courses. These teacher training centres are still in their formative stage but will be very important institutions in view of the stress laid on non-degree programmes in agriculture and the 10 + 2 + 3 pattern of education as national policy.

Summer institutes

Teacher training at the university/college level has been the missing link. While a number of seminars, workshops and conferences have been organized on an ad hoc basis in different parts of the country for college/university teachers, the summer institutes sponsored by ICAR since 1965 have had a major influence in updating teachers. The summer institutes have been conducted by the agricultural universities, colleges and institutes in specialized areas where the institutions concerned have established a reputation in the field of teaching and research. Though the programmes of such summer institutes should encompass both technical subject-matter and teaching technique, experience has shown that teacher training programmes in the past have to a great extent concentrated on the subject-matter. A proper balance of the two should be the consistent design of the summer institute programmes.

For each summer institute, a programme director is appointed and a group of faculty members of the institution are involved. In addition, outside resource persons are also called on. Normally, the duration of the summer institute is one month, and twenty to thirty teacher trainees benefit from each course. The summer institutes so far organized for agricultural teachers are shown in Table 19.

TABLE 19. Summer institutes organized for agricultural teachers during 1967/77 in different disciplines

Year	Number of courses organized	Number of teachers trained
1967	1	34
1968	5	120
1969	6	150
1970	10	198
1971	10	246
1972	15	335
1973	17	419
1974	21	436
1975	14	300
1976	16	344
1977	23	500
TOTAL	138	3 082

Integration of teaching, research and extension

Integrated functioning of research, teaching and extension is the most fundamental concept of an agricultural university. These three functions—creation of knowledge, imparting of knowledge and skill, and dissemination of the applied knowledge and skills are inherently interwoven, and their functional interaction is imperative in order to maximize their effectiveness. Training, at present part and parcel of the extension function, is emerging as a potentially major area of activity in the universities and hence requires added attention and specific treatment. In fact, this should be considered as the fourth major function of the agricultural university. Integrating these functions has not been an easy matter; a continued effort, therefore, is essential.

Earlier, the colleges were mainly responsible for teaching. The research institutions were under the jurisdiction of the state/central government, and, as explained earlier, the university extension function was limited to an extension education role, rather than an extension service role. A clear demarcation of functions and relationships between the university and state government had to be made, and this has taken relatively longer than was expected. Some state governments have yet to transfer full research responsibility and the related institutions to their respective agricultural universities.

Integration has to take place through administrative policy decisions and at the individual level. At first, university staff were expected to take an equal part in all the above functions, but it was soon realized that a good teacher may not necessarily be a good researcher nor a good research scholar an effective extension agent and vice versa. Thus, this principle is gradually being recognized by university authorities and functions are being assigned to a great extent in accordance with the interest, aptitude and proficiency of the staff. Nevertheless the universities have been somewhat handicapped by the lack of adequate staff.

The agricultural university design was introduced into the already existing educational infrastructure with a few exceptions. Of the twenty-one agricultural universities, only six are mono-campus type and the remaining fifteen have multi-campuses. Naturally, the multi-campus structure has made integration relatively difficult. Further, most staff members lacked the appropriate insight and working experience to appreciate the intricacies and implications of integrated working. This has resulted in a slow development of integration.

In some universities, clarification is needed as to the role and relationship between the dean of instruction, the director of research, the director of extension, the associate deans of the campuses and the heads of departments. Experience has shown that integration should first take place at the individual level, where the individual teacher and researcher, while appreciating the total function of the university, contributes directly to his interest area and indirectly to the other areas. His direct or indirect association with all university functions is essential if his work is to be fully relevant. Based on this premise, the heads of departments should co-ordinate and integrate the work of the departments, and should develop functional linkages with other departments, wherever necessary. The associate deans, deans and directors should play a similar role at their respective levels and should provide the policy framework most suited to the effective integration of university functions.

Agricultural training for development

Farmer training is an important input in the process of agricultural development. The census report of 1971 revealed that there were, on an average, 210 potential farmers/agricultural workers per village, 380,000 per district and 130 million in the country as a whole. Skill training on a continuous basis for these practising farmers and workers is a gigantic task. The scope of training has further been enlarged in view of new technical breakthroughs and the increasing sophistication of the farming systems. The training of the trainers/teachers is yet another task which has to be reckoned with.

A considerable gap exists between the available technologies and their adoption by farmers. The inadequacy of skill training of both trainers and trainees has been cited as one of the basic limiting factors in the rapid transfer of agricultural technology.

Training of development staff

Directorate of Extension, Ministry of Agriculture

Since the inception of the Community Development Programme (CDP) in 1952, the need for subject-matter support to the extension workers at different levels has been appreciated. Hence subject-matter-oriented refresher training courses in agriculture and animal husbandry for extension staff was initiated in selected agricultural and veterinary colleges in 1958. Later the Extension Education Institute, Nilokheri, started organizing integrated training courses in subject-matter and extension techniques for various extension officers. Up to 1975 the institute had organized 194 in-service training courses for 3,709 extension functionaries. Table 20 on the following page presents examples of courses offered in the Nilokheri, Anand and Rajendranagar institutes.

Since the early 1960s, a number of development programmes have been launched to step up agricultural and allied production, e.g. the Intensive Agricultural Area Programme (IAAP) in 1964/65, Intensive Cattle Development Project (ICDP) in 1964/65, High-yielding Varieties Programme (HYVP) in 1966/67, Small Farmers Development Agency (SFDA) in 1970/71, etc. To train the senior officers of these and other projects, the Directorate of Extension (Ministry of Agriculture) designed short-term staff courses in agriculture in 1966. These courses were organized in association with the agricultural universities, research institutes and colleges. The duration of courses varied from four to ten days with twenty-five to thirty participants per course.

By 1974/75, sixty courses on crop production technology and twenty-two

TABLE 20. Training courses organized at extension education institutes up to 1975

Course	Duration	Extension education institutes							
		Nilokheri		Anand		Rajendranagar		Total	
		Number of courses	Persons trained	Number of courses	Persons trained	Number of courses	Persons trained	Number of courses	Persons trained
Integrated training course in subject-matter and extension techniques for extension officers (agriculture and animal husbandry)	6 weeks	18	356	20	405	17	358	55	1 119
Course for village artisans and farm mechanics	1 year	17	256	—	—	—	—	17	256
Refresher course for agricultural engineers	4 weeks	9	111	—	—	—	—	9	111
Refresher course for village-level workers	2 months	48	1 028	—	—	—	—	48	1 028
Training course in agriculture credit for extension officers working in SFDA/MFAL areas	2 weeks	3	54	2	53	2	51	7	158
Training course in agriculture credit management for district-level officers working in SFDA/MFAL areas	12 days	1	24	1	13	1	19	3	56
Workshop in agricultural administration and management for district-level officers of agriculture departments	6 days	1	13	—	—	—	—	1	13
Miscellaneous courses and courses for other categories of extension personnel in applied nutrition research methodology, food demonstration	3 days	32	509	7	161	15	298	54	968
TOTAL									3 709

courses on animal husbandry had been organized, benefiting 1,427 and 357 officers respectively.

Directorate of extension education of the agricultural university

Ever since the creation of agricultural universities, their directorates of extension education have been playing a decisive role in organizing training programmes for different levels of agricultural development staff. Courses of different duration have been organized on a regular as well as on an ad hoc basis. For example the Gobind Ballabh Pant University of Agriculture and Technology has, since 1963, organized a number of training courses for staff development. As an illustration, the courses organized during 1977/78 are shown in Table 21.

TABLE 21. Staff training courses organized at Gobind Ballabh Pant University of Agriculture and Technology during 1977/78

Subject number	Development staff	Duration in days	Total staff trained
1	District agricultural officers	28	4
2	Extension officers of private agencies	11	9
3	Extension training officers, subject-matter specialists, etc.	2	15
4	Staff of Indian and State Administrative Service (IAS/PCS)	14	37
5	Deputy directors of agriculture and district agricultural officers	3	26
6	Sub-divisional magistrates	4	23
7	Subject-matter specialists and additional district agricultural officers	2	45
8	Seed development officers	30	4
9	District development officers	6	11
10	Technical staff in agriculture	30	12
11	Branch managers of banks	10	35
12	District magistrates	5	6
13	PCS probationers	13	34
14	Technical staff of state in agriculture	6	22
15	Agricultural attaché to IAS officers	6	24
16	Assistant development officers	6	18
17	Other extension staff	1-30	511

The directorate of extension, Punjab Agricultural University, has, since 1963, organized a large number of training courses for officers of development departments who are directly or indirectly concerned with agricultural production. Important courses for officers directly related to agricultural development are: (a) agricultural officers workshop for *kharif* and *rabi* crops, and (b) refresher training courses for the various categories of extension personnel of the departments of agriculture, animal husbandry, soil conservation, co-operation and development. During 1976/77, 919 such development staff were trained for three to eight days. Courses for officers indirectly concerned with agricultural development are: (a) gardening and landscaping for schoolteachers; (b) agricultural production and management for Border Security Force and army officers; (c) agriculture and finance for bank officers; and (d) training courses under the Applied Nutrition Programme (ANP) for block development officers, ladies circle supervisors, and village-level workers. The university trained 254 such in-service staff during 1976/77.

The remaining nineteen agricultural universities have been training development staff from the respective state governments to a greater or lesser extent depending upon the strength of their directorates of extension education. The directorates of extension of all the twenty-one agricultural universities in the country are being assisted by ICAR to strengthen their training, advisory and communication services. The training contributions of these universities have been considerable, and enlarged programmes are expected in the future.

ICAR institutes and staff college

In addition to supporting agricultural research, education and training in the country mainly through agricultural universities, ICAR directly controls 31 research institutes, 52 All India co-ordinated research projects, 142 research centres/sub-stations, one central staff college and 7 teacher training centres. While the main concerns of the research institutes and research centres/sub-stations, and also the All India co-ordinated research projects, have been to generate knowledge through research studies, a few are also committed to the higher education and training of research and development staff. Indeed training is gaining momentum in all the institutes/centres in vew of its importance as a means of transferring technology for rural development. The Central Staff College for Agriculture is entirely devoted to the training of research and development staff; teacher training centres to that of teachers/trainers and development staff; and KVKs to the training of field-level extension functionaries as well as to that of farmers.

The extension wing of the National Dairy Research Institute, for example, organizes some regular courses all the year round in addition to ad hoc courses. Table 22 provides an overall picture of the training courses which were organized in the year 1977/78.

The Central Soil Salinity Research Institute, Karnal, has been organizing ad hoc courses, such as the reclamation of alkali soils, for senior officers and inspectors of the state government, and the management of salt-affected soils for agronomists and soil scientists working in the fertilizer industries. The Indian Grassland and Fodder Research Institute, Jhansi, has organized courses in silvipasture and grassland production management, fodder production, range management and its utilization, and grass and legume seed production for development staff of the state government. Similarly the Central Soil and Water Conservation Research and Training Institute, Dehradun, has organized regular courses for gazetted officers and non-gazetted assistants working in the field of soil and water conservation. Since 1955/56 it has trained 1,101 gazetted officers and 3,560 non-gazetted assistants. By now most of the institutes offer some training programmes to development staff. However, the smaller research centres/sub-stations have yet to develop training support to development.

The Central Plantation Crops Research Institute, Kasaragod, has been organizing short-term regular training courses for development staff of the state governments on various aspects of plantation crops—coconut, areca nut, cacao, cashew, oil palm and spices. Table 23 gives a résumé of staff training between 1972 and 1978.

TABLE 22. Training course organized by NDRA, Karnal, during 1978 for Dairy Development ment Programme staff in various fields

Course	Duration in days	Number of participants per course	Total number of participants
Dairy extension practices	10	10	63
Management of dairy animals	30	15	29
Improved dairy farming practices	30	20	64
Dairy farming technology	10	18	18
Condensed and dried milk products	30	3	3
Butter-making	21	3	3
Cheese-making	30	3	3
Flavoured/sterilized milk	15	5	5
Fermented milk	15	4	4
Ice-cream-making	21	10	10
Ghee-making	15	2	2
Management of Dairy Development Programme	30	4	4
Bank officers' special course	7	33	33
Dairy managers' business	30	10	10
TOTAL			251

TABLE 23. Personnel trained in plantation crops

State	1972	1973	1974	1975	1976	1977	1978	Total up to 30 October 1978
Andamans	—	—	—	—	—	—	12	12
Andhra Pradesh	7	7	3	3	7	—	—	27
Assam	—	1	—	—	1	—	—	2
Goa	—	—	—	1	—	2	—	3
Gujarat	3	4	—	—	1	—	1	9
Karnataka	8	10	4	5	—	—	—	27
Kerala	11	10	9	15	3	58	13	119
Lakhadweep	—	—	1	—	—	5	—	6
Maharashtra	4	5	—	—	1	—	—	10
Mizoram	—	2	—	—	2	—	—	4
Orissa	1	3	—	3	4	—	—	11
Pondicherry	7	1	—	2	—	—	—	10
Rajasthan	—	—	—	—	1	—	—	1
Tamil Nadu	4	11	13	4	4	—	—	36
Tripura	—	1	—	—	—	—	—	1
West Bengal	—	—	—	—	1	—	—	1
North-east Complex	—	—	—	—	—	12	9	21
Foreign countries	—	—	3	—	—	5	4	12
TOTAL	45	55	33	33	25	52	39	312

In view of the necessity of staff training, both for institutional management and for the supervision and guidance of development programmes, the Central Staff College for Agriculture was established at Hyderabad in 1976. Its main task is to organize advanced short training courses in planning, management, evaluation and monitoring of institutions and projects for senior-level research and development functionaries. So far eight batches of agricultural research probationers have been trained in the two- to three-month foundation courses in research management, totalling 752 trainees. In 1978 the college also initiated training for senior scientists in a two-week course attended by twenty-five scientists from different institutes.

State institutes of community development

The concept of democratic decentralization emerged in 1959. Under this system, non-official bodies—Village Panchayat, Panchayat Samiti and Zila Parishad—were created at village, block and district levels respectively. Thus the involvement of non-official elements in community development was ensured.

Study and orientation training centres were established in all the states in order to train officials and non-official participants in community development programmes (CDP) and to equip them for a larger contribution and more effective participation. In some states, this institution was called a *panchayati raj*[1] training centre. Recently these centres have been designated state institutes of community development. More than a dozen are now functioning and offer pre-service and in-service training courses for community-development workers, and orientation training courses for the non-official elements.

Both on-campus and peripatetic training programmes of varying duration are organized on the basis of needs. The subjects included in the training programmes are: *panchayati raj* system, village planning, agricultural production plans, community-development working, co-operatives in developments, rural institutions, etc.

National Institute of Rural Development

The Central Institute of Study and Research was set up in 1958 at Mussoorie, Uttar Pradesh, for training senior community-development functionaries. A trainers' training institute came into being during the same period in order to train regional and state-level trainers of institutions such as the study and orientation training centres, *gram sevak* training centres, etc. Following the introduction of *panchayati raj*, this institute was renamed the Institute for Instruction in Community Development. In due course it was felt that the above two institutions could serve a more useful purpose if they were merged. Thus, the National Institute of Community Development was created in 1962. In 1977, in line with the policy of government to stress rural development, the institute was again renamed and became the National Institute of Rural Development.

To start with, training programmes were conducted for senior officers involving formulation and implementation of extension programmes. The courses have since been diversified in order to meet the training demands of various categories of rural development functionaries. During 1966/77, the institute organized

1. Elected village council.

157 training courses benefiting 4,331 participants. The courses encompassed such areas as village and block-level planning, formulation of rural development projects, project implementation, monitoring and evaluation, rural institutions in development, management in rural development administration, etc.

Training of farmers

Universities and institutes

As already noted, the directorates of extension of the agricultural universities have provided farmer training on a limited scale for selected community development blocks/districts. Punjab and Haryana agricultural universities have established training schools for farmers, especially young farmers. In both these states, subject-matter specialists of the universities have been stationed at district level in order to organize short training programmes and to provide advisory services for farmers, in collaboration with district development staff of the state governments. Such organized systems have yet to be adopted by other agricultural universities, though they have organized some training programmes for farmers of the community-development blocks in their areas.

For example, the Punjab Agricultural University has provided ten days of special training courses for farmers in the areas of farm machinery, poultry farming, dairy farming, pig raising, fruit and vegetable preservation, etc. The farmers are given free board and lodging during the course. Since 1971 the university has started correspondence courses in subjects related to agriculture for educated farmers (matriculates and above) who are actively engaged in farming. The duration of the course is one year and the medium of training is the local language (Punjabi). A similar course was initiated for farm women in 1973 in subjects related to home science. The course is open to Punjab farm women who have at least completed primary school. Training courses have also been organized for members and officers of *yuvak dals* (youth clubs), and also for participating local advisers and heads of rural institutions. The university also runs a farmers' training centre on behalf of the Government of India. Table 24 shows the training courses organized by Punjab Agricultural University during 1976/77.

In addition, during the same year, under the farmers' training and education programmes, 662 courses of one to two days' duration were organized, in which 11,847 farmers and farm women were given training. The university, under the youth development programme, had also organized fifty-two training courses of one to two days' duration and four courses of three months' duration benefiting 2,131 young farmers.

ICAR institutes, being basically committed to research, have not given much attention to farmers' training, but it has now been realized that they must also assume some extension functions, including training. The national research institutes (National Dairy Research Institute (NDRI), Indian Agricultural Research Institute (IARI), and Indian Veterinary Research Institute (IVRI)) have already established extension divisions. NDRI has been organizing training of one to two weeks' duration in the areas of feed and fodder management, milk and milk products, etc. for farmers. IARI has been regularly organizing short training programmes for Delhi farmers at the start of both *rabi* and *kharif* seasons. They

TABLE 24. Training programmes organized by the Punjab Agricultural University during 1976/77

Course	Duration	Trainees per batch	Number of batches in the year	Total farmers trained
Specialized training courses in agricultural field:				
Tractor and farm machinery	10 days	22	5	108
Dairy farming	10 days	22	4	87
Poultry farming	10 days	30	2	59
Pig raising	10 days	29	. 1	29
Bee-keeping	15 days	30	2	59
TOTAL			16	342
Correspondence courses (integrated) programme:				
Agriculture (for farmers)	1 year	314	1	314
Home management and family life (for farm women)	1 year	178	1	178
Farm power machinery maintenance and operations (for farmers)	1 year	48	1	48
Special course in agriculture/home management and family life (for small and marginal farmers/ farm women)	1 year	261	1	261
TOTAL		801	4	801

have also organized leadership training courses for farm youth and schoolteachers who have been associated with youth club work and rural development projects. The Indian Lac Research Institute, Ranchi, has conducted courses on improved methods of lac cultivation and the uses of lac for farmers and entrepreneurs, respectively. Many other courses have now been initiated by various research institutes on both an ad hoc and regular basis for farmers, fishermen, farm women, etc.

Farmers' training centres

With the introduction of the High-yielding Varieties Programme (HYVP), an acute need for farmer training was felt. Consequently, a centrally sponsored scheme for farmers' training and education was launched by the Ministry of Agriculture, Government of India. This scheme envisaged organizing farmer training duly supported by farm radio broadcasts and the functional literacy programmes. The objectives of the training programme were:

To provide concurrent technical know-how relating to the high-yielding varieties and multiple cropping programmes, and the scientific use of agricultural inputs such as seeds, fertilizers, pesticides, water, etc.

To create a core of progressive farm leaders.

To encourage farmers to develop an interest in seeking guidance from extension personnel and agricultural scientists regarding the problems facing them in their fields.

To assist farmers by imparting knowledge regarding the resources, availability of various agricultural inputs; mode of securing assistance and credit, etc.

The centre is headed by the district training officer and supported by six other trainers—training officers (male and female), radio contact officer, management specialist and demonstrators (male and female).

The work done up to 1978 appears in Table 25.

TABLE 25. Training courses and discussion groups organized up to March 1978 under farmers' training programme

Subject number	Training courses and discussion groups organized	Number of courses/ discussion groups	Number of participants
1	Specialized short courses for farmers	10 187	294 131
2	Specialized short courses for farm women	6 034	176 198
3	Specialized courses for convenors of discussion groups	4 072	59 728
4	Production-cum-demonstration training camps	80 525	2 838 207
5	Farmers discussion groups	27 822	552 805
6	Farm women discussion groups	7 150	146 845
7	Three-month training courses for young farmers	73	2 025
	TOTAL	135 863	4 069 939

Source: Progress Report—Farmers' Training and Education Programme (1977/78), New Delhi, Directorate of Extension Ministry of Agriculture and Irrigation, Government of India.

Farmer training centres organize twenty specialized five-day on-campus training courses in a year, in which twenty-five trainees are briefed on the latest technology in crop production and livestock production; and in home science and management for farm women. In addition, the training centres conduct 100 production-cum-demonstration camps of one day's duration during the year on either a demonstration plot or the fields of progressive farmers. The convenors (125 in number) of Radio Rural Forums *(Charcha Mandals)*, are also given training of three days' duration in order to keep them abreast of the latest developments in agricultural technology.

Panchayati raj training centres

Within the concept of democratic decentralization, an attempt was made to encourage peoples' participation in decision-making and development activities through *gram panchayats* (village councils), *panchayat samities* (block councils) and *lila parishads* (district councils).

For the training of elected local leaders and the secretaries of these councils, the *panchayati raj* training centres, now known as state institutes of rural developments were established in different states. Two types of course have been organized in these institutes—institutional training for a week and peripatetic training camps for three days. Subjects include the concept of democratic decentralization; functioning of *panchayati raj* bodies; village-, block- and district-level planning for agriculture and allied development; role of village institutions in development; organization of *yuvak mandals* (youth clubs), *bal mandals* (children's play-centres) and *mahila mandals*; working of co-operatives, health and sanitation; and family planning, etc. At present there are over a dozen state institutes of rural development in the country.

Krishi Vigyan Kendras

Farmers' training programmes which in the past were organized by various institutions have suffered because of (a) weak subject-matter support; (b) academic approaches and methods of training; (c) absence of facilities for practical training; (d) training programmes unrelated to immediate needs; (e) stress on quantity rather than quality; and (f) limited financial support for training infrastructure. To overcome these barriers to agricultural production, the scheme of Krishi Vigyan Kendra (KVK) was first initiated in 1974 by ICAR and expanded after 1976.

Concepts and objectives

KVK is a grass-root-level institution designed to impart need-based and skill-oriented short- and long-term vocational training courses. Dr Mohan Sinha Mehta's committee laid down the concepts of KVK as follows:

KVK will impart learning through work experience and hence will be concerned with technical literacy, the acquisition of which does not necessarily require as a pre-condition the ability to read and write.

KVK will impart training only to those extension workers who are already employed or to practising farmers and fishermen. In other words, it will cater to the needs of those who are employed or those who wish to be self-employed.

There will be no uniform syllabus for KVK. The syllabus and programme of each will be tailored according to the felt needs, natural resources and the potential for agricultural growth in that particular area.

KVK is based on three fundamental principles; first, agricultural production is the prime goal; second, work experience (learning by doing) is the main method of imparting training and education; and third, priority is given to the weaker sections of the rural population. The idea is to influence the production system through social justice, the starting-point being the most needy and deserving section of the society—the weaker sections, tribal farmers, small and marginal farmers, agricultural labourers, drought- and flood-affected farmers, etc.

A majority of the trainees will be either school drop-outs or illiterate farmers. Some general education is, therefore, also imparted to make them not only good farmers but also alert citizens.

In view of the sophistication of modern farming, as also the low level of literacy in rural areas, skill training of the farming communities has become imperative. In each KVK, therefore, practical training facilities, such as a farm for 'earn-while-you-learn' projects, an agricultural workshop, a small dairy unit, poultry farm, fish ponds, etc. are being meticulously developed. The work experience is never in the form of labour-oriented activities. It is an educational process where the trainees learn operative skills and develop confidence in tackling their farm problems. This skill development takes place in the context of a fair understanding of the scientific principles and processes involved.

No certificate or diploma is given for either short or long training courses in order to attract practising farmers, or those youths who wish to be self-employed. The programmes are composite, covering all agricultural and allied aspects. The intention is to develop each KVK as a rural training centre, including cottage industries, consistent with the requirements of integrated rural development.

The scheme provides for a dozen scientific staff belonging to different fields of specialization such as agronomy, horticulture, livestock production, home science, fisheries and extension education. This is, in fact, a core staff; in addition specialized resource people or visiting trainers are invited on an ad hoc basis for specific short-term courses. Based on local needs, even the progressive farmers, artisans, fishermen, etc. can be invited as paid visiting trainers. In appointing the staff, noted earlier, the quality of field and training experience is given heavy weightage in relation to academic qualifications and achievements. In addition, the necessary supporting office staff have also been provided.

KVKs are being linked with the research institutes and agricultural universities for technical back-stopping. An effort is being made to develop functional linkages with the state development departments/projects and other agencies engaged in rural development.

TABLE 26. Training courses organized and number of farmers trained during 1977/78 at KVKs

Subject number	Area of training	Number of courses organized	Number of trainees trained
1	Crop husbandry	235	6 430
2	Livestock production	214	4 219
3	Fisheries	17	350
4	Horticulture	42	930
5	Home science	50	1 004
	TOTAL	558	12 933

Source: Proceedings of the Second All India Workshop on Krishi Vigyan Kendra, ICAR, 1978.

Agricultural training through non-formal programmes

India being an agricultural country, any social education efforts and non-formal educational programmes should touch directly or indirectly on the agricultural and allied sectors. Thus, by being involved in agricultural-based activities, young men and women are helped to learn the rudiments of scientific agriculture. These programmes are mainly implemented by the departments of social welfare, departments of rural development, the National Council of Educational Research and Training, educational institutions and voluntary organizations.

National Cadet Corps (NCC). The NCC programme is a well-organized venture for training young men and women with the object of developing character, comradeship, a desire to serve the nation, and leadership ability. As part of their total training, they are required to undertake a variety of social activities, some of which are focused on agriculture—digging wells for irrigation, deepening of irrigation canals, rat control, afforestation, etc. The scheme is voluntary. Management and expenses are shared by the central and state governments.

There are two levels of cadets—junior and senior; the former for schoolboys and girls, and the latter for college-level students. At present there are over 625,000 junior-level and about 350,000 senior-level cadets on roll. In view of the increased stress on rural and agricultural development, NCC cadets now are expected to be involved in agricultural projects on a relatively larger scale.

Planning forums. This scheme was first initiated by the Planning Commission in 1955, but was transferred to the Ministry of Education in 1968. The planning forums aim at: (a) creating an awareness among the student community of the need for planned development; (b) developing plan consciousness among educated youth in particular and through them the general public; and (c) involving students in the national building effort right from the planning stage.

The planning forums function through universities/colleges and other approved educational institutions. The activities of the National Social Service (NSS) and the planning forums are complementary. The Central Ministry of Education and the state governments share project costs on a 60 : 40 basis. The co-ordination committees at the university, state and central levels monitor the progress and performance of the planning forums.

The activities of the forums include organizing discussions and debates on planning, planned development and related areas; running plan information centres; conducting socio-economic surveys; providing feedback to development departments/institutions; and arranging seminars and educational tours to development areas. All their rural programmes include agricultural aspects; the planning forums held in agricultural universities/colleges devote all their programmes to agricultural and rural development. There are at present 1,000 planning forums functioning in various institutions.

Rural youth clubs. The development of the community through youth participation and the training of youth was recognized in the early extension efforts (1920s) in the country. Nevertheless, the development of rural youth clubs in an organized manner was taken up after the Annual Conference on Community Development held at Mysore Karnataka, in 1959. The object laid down for youth club work was to help farm youth to become better farmers and enlightened citizens through individual or group projects/activities.

The village youth clubs were organized for both boys and girls as an integral part of the activities of community development programmes. Youth club members were expected to undertake group or individual economic projects in their leisure time as a basis for learning as well as earning. Popular projects have been kitchen gardening, poultry-keeping, calf-rearing and improved crop cultivation for boys; and knitting and sewing, foods and nutrition, animal feed and care, goat-rearing, etc. for girls. The clubs are organized by village-level workers, social education organizers and social workers in association with local institutions and village leaders.

In 1967/68, there were 1.16 million youth clubs in 4,303 community-development blocks with a membership of 23.93 million rural youths (boys). During the same year there were 58,200 youth clubs for girls *(mahila mandals)* with 1.39 million members.

Radio Rural Forums. Under the scheme for farmer training and education, farmers' discussion groups *(churcha mandals)* or Radio Rural Forums were organized. Participating farmers are organized into radio listening groups, are expected to listen to the agricultural programmes on the radio regularly and then to hold discussions among themselves. The groups are provided with subsidized radio sets as an incentive. Such groups are providing useful feedback to All India Radio as well as to the farmers' training centres.

Special training courses are organized for the convenors of the *churcha mandals.* Up to 1978, 4,072 such courses were organized for 59,728 participants of the 27,822 farmers' discussion groups and 7,150 farm women's discussion groups having 552,805 and 146,845 members respectively.

Farmers' functional literacy. Literacy and development are being organically linked through the Farmers Functional Literacy Programme (FFLP) *(Kisan Saksharta Yojna)* of the Ministry of Education, Government of India. It was launched in the year 1967/68.

FFLP has an integrated three-dimensional approach under three union ministries: the Ministry of Education and Social Welfare, the Ministry of Agriculture and Irrigation, and the Ministry of Information and Broadcasting. These have developed linkages between functional literacy, farmers' training and farm broadcasting. At the state level the State Department of Education looks after FFLP in collaboration with other development departments of the government. There are co-ordination committees at the central, state and district levels.

The District Education Officer has overall charge of the project. He is assisted by a full-time literacy project officer for execution and supervision of the programmes. At the village level part-time supervisors (one for ten classes) provide further assistance. The classes are taken by part-time instructors who are normally schoolteachers and educated farmers and who receive a nominal honorarium. By 1977/78 over seventy-five districts, each with sixty to seventy village centres, had been covered under the scheme, benefiting about 400,000 farmers.

National Service Scheme (NSS). NSS was started in the year 1969/70 for college students. The scheme aimed at: (a) helping to bridge the gulf between the élite and the general public; (b) bringing the educational institutions into closer contact with the community; (c) developing in youth the realities of life and living; (d) enriching their experiences and developing respect for the dignity of labour; and (e) preparing youth to serve as an instrument of social change.

Through this scheme students in the first two years of the three-year degree course are organized for participation in community service activities in villages or in selected urban localities, mainly by organizing camps in the selected area/ villages. The programmes include creating minor irrigation facilities for farmers —digging or deepening wells and tanks, disseminating improved methods of cultivation especially among the smaller farmers, arranging for credit facilities, educating farmers about land reform measures and social legislations, etc. On occasion, they have organized special campaigns such as 'Youth Against Famine', 'Youth Against Dirt and Disease', 'Youth for Afforestation', etc. Recently NSS workers have been concentrating their efforts on the task of 'rural reconstruction' by adopting villages.

At the outset thirty-eight universities participated in this programme, involving 40,000 students. By now, 117 universities/institutions of higher learning are enlisted in NSS. There are 2,472 colleges implementing this programme and 270,000 students are involved.

Non-formal education for children. In spite of the importance being given to universalization of primary education, only 83 per cent of children in the 6-11 age-group and about 37 per cent in the 11-14 age-group are attending school. Of these, almost 60 per cent drop out before completing class V and 75 per cent before completing class VIII. These drop-outs later relapse into illiteracy.

A pilot project to organize a non-formal educational programme for these school drop-outs was started in 1975 by NCERT at Bhumiadhar (Uttar Pradesh). The main purpose was to provide some educational opportunity right in the vicinity so that they could join the regular school stream in class VI or class IX. It was soon realized that such a programme could not succeed unless a comprehensive community development programme was initiated in which both parents and children could be involved. Thus, a few work experience oriented agricultural and allied activities, such as demonstrations of the cultivation of 'man-made-wheat', cultivation of mushrooms, making compost pits, etc., were introduced together with other activities such as ceramic work, dyeing, etc.

The project at Bhumiadhar was successfully completed for forty children of the 6-14 age-group but a large number of others benefited through organized educational activities.

Nehru Yuvak Kendra (NYK). Based on the recommendations of the National Advisory Board on Youth (1970), the Nehru Yuvak Kendra (Nehru Youth Centre) scheme was initiated in 1972. The object was to organize various informal and non-formal educational programmes for out-of-school youth.

NYKs are being set up at the district level as a nucleus for educating and mobilizing youth. The Kendra has a youth co-ordinator with an accountant and a part-time helper. The co-ordinator enlists the co-operation of local and outside institutions/agencies. The Kendra programmes include organizing vocational and technical training, village youth clubs, family life education, reading-rooms and library services, social services such as work for better rural environment, work for slum improvement, construction and repair of roads, etc., sports and games, and recreation and cultural activities.

The Ministry of Education and Social Welfare sponsors this scheme, which is implemented by state governments. Co-ordination committees review the work and provide guidance.

By 1978, 235 NYKs had been created.

Soil Health Care. This programme envisaged training in soil testing for chemistry teachers of general science colleges who, in their turn, were to train chemistry students to take up soil health care in the adopted villages. The scheme was initiated in 1977 and ICAR, UGC and the Ministry of Education collaborated. The aim is to involve college youth participating in the NSS programme as soil health care workers in rural areas.

Sixty science colleges in twelve states where NSS operates have been chosen under the scheme to start with. ICAR has agreed to provide funds for training sixty selected chemistry teachers in agricultural universities, while UGC is providing soil-testing equipment and chemicals. Each of the trained teachers will train twenty-five students from his respective college as soil health workers.

Training of the teachers has already been completed and the students are now being trained. Each college is expected to prepare, with the help of soil health workers, soil maps for at least two adopted villages. The scheme has a significant potential for multiplication.

Terminology
in agricultural education
and training

extension functionary *(prashar karya-
karta)*, 59, 60
extension service department *(prashar
seva vibhag)*, 21
extension training *(prashar prashikshan)*, 10
extension training centre *(prashar
prashikshan kendra)*, 16, 17

Farm Science Centre *(Krishi Vigyan
Kendra)*, 10, 22
Farmers' Functional Literacy Project
(FFLP) *(Kisan Saksharta Yojna)*,
23, 73
farmers' training and education *(kisan
prashi kshan and shiksha)*, 67
farmers' training centre (FTC) *Krishak
Prasikshan Kendra)*, 54
fisheries education and training *(matasya shiksha aur prashikshan)*, 17
Fisheries Extension Training Centre
*(Machhdi Plan Vistar Prashikshan
Kendra)*, 51
fisheries high school *(matasya uchcha
vidyalaya)*, 15
fisheries operatives *(matasya prayojak)*,
13
five-year plan *(panch warsheeya hoja-
na)*, 23
forest education *(van shiksha)*, 11
forest rangers *(vanpal)*, 11, 16
forest rangers' college *(vanpal naha vidyalaya)*, 11, 35
forest school *(vidyalaya)*, 11
formal education *(aupeharik shiksha)*,
9

general universities *(sadharan vishwavidyalaya)*, 19
government farm *(sarkari prashetra)*,
13
graduate programme *(sanatak karya-
kram)*, 40-5
gram sevik training centre *(gram sevik
prashikshan kendra)*, 15
*gram sevika (woman village level
worker)*, 15, 34

health extension educator *(swasthya
vistar shikshak)*, 34

health visitor *(swasthya nirikshak)*, 34
higher agricultural education *(uchcha
krishi shiksha)*, 11, 17, 37
higher education *(uchchatar shiksha)*, 9
High-yielding Varieties Programme
(HYVP) *(Adhik Upajwali Kismon
Ka-Karyakram)*, 61
home science *(grih vigyan)*, 9, 14, 33

ICAR, Institutes and Staff College
*(Bhartiya Krishi Anusandhan Pa-
rishad-Aur Adhikari Mahavidya-
laya)*, 64, 67
Imperial Agricultural Research Insti-
tute (IARI) *(Shahi Krishi Anusan-
dhan-Sansthan)*, 12
Imperial Council of Agricultural Re-
search *(Shahi Krishi Anusandhan-
Parishad)*, 12
Imperial Dairy Research Institute
*(Shahi Dugdhashala Anusandhan-
Sansthan)*, 77
Indian Agricultural Research Institute
(IARI) *(Bhartiya Krishi Anusan-
dhan-Sansthan)*, 13, 19, 37
Indian Constitution *(Bharatiya Sam-
vidhan)*, 9, 13
Indian Council of Agricultural Educa-
tion (ICAE) *(Bhartiya Krishi
Shiksha-Parishad)*, 19
Indian Council of Agricultural Re-
search (ICAR) *(Bharatiya Krishi
Anushandhan-Parishad)*, 10, 17,
19, 21, 23, 64
Indian Dairy Diploma (IDD) *(Bhartiya
Dugdhashala Adhikar-Patra)*, 11,
16, 31
Indian Forest College *(Bhartiya Van
Mahavidyalaya)*, 19
Indian Forest Service Probationers
*(Bhartiya Van Sewa Parivikshar-
thi)*, 51
Indian Grassland and Fodder Research
Institute *(Bhartiya Ghas Tatha-
Chara Anusandhan-Sansthan)*, 64
Indian Institute of Management
(Bhartiya Prabandha Sansthan), 51
Indian Lac Research Institute *(Bhar-
tiya Lakh Anusandhan Sansthan)*,
68

Indian Veterinary Research Institute (IVRI) *(Bhartiya Pashu Chikisha-Anusandhan Sansthan)*, 11

Indo-American Team *(Bhartiya-American Dal)*, 17, 19

induction training *(sewa pravesh prashikshan)*, 21

informal education *(anaupcharik shiksha)*, 9

Inland Fisheries Training Unit (IFTU) *(Untahsthaliya Matasya Prashikshan-Kendra)*, 19

in-service training *(seva-antargat prashikshan)*, 21

Institute of Agricultural Research Statistics (IARS) *(Krishi Anusandhan Sankhikee Sansthan)*, 19

Institute of Applied Manpower Research *(Vyavahark Sanbal Anusandhan-Sansthan)*, 49

integrated diploma course *(samakalit adhikarpatra adhyayan)*, 16

Integrated Fisheries Project *(Samakalit Mabhali Palan Prayojana)*, 23

integrated rural development *(samakalit gram vikas)*, 15, 52

integration of teaching, research and extension *(shikshan, anusandhan-tatha vistar ka samakalan)*, 9, 59

Intensive Agricultural Area Programme (IAAP) *(Saghan Krishi Kshetra-Karyakrama)*, 22, 61

Intensive Agricultural District Programme (IADP) *(Saghan Krishi Zila-Karyakrama)*, 22

Intensive Cattle Development Project (ICDP) *(Saghan Gopashu Vikas Prayojna)*, 61

intermediate education in agriculture *(krishi me uchchattar madhyamik-shiksha)*, 10

junior basic school *(kanishta buniyadi vidyalaya)*, 13, 24

Krishi Vigyan Kendra (KVK) (Farm Science Centre), 10, 22, 64, 70

lower primary *(nimna prathmik)*, 77

medical education *(vaidyak shiksha)*, 9

middle school *(madhyamik vidyalaya)*, 13, 24

Ministry of Education and Social Welfare *(Shiksha Aur Samaj Kalyan-Mantralya)*, 9, 23, 27

Mudaliar Commission *(Mudaliar Ayog)*, 14

multi-purpose high school *(bahuddeshiya uchcha vidyalaya)*, 14, 25

National Advisory Board on Youth *(Rashtriya Yuva Salahkar Parishad)*, 23

National Board of Adult Education *(Rashtriya Praudra Shiksha Mandal)*, 10

National Cadet Corps *(Rashtriya Sainik Dal)*, 23, 72

National Commission on Agriculture *(Rashtriya Krishi Ayoga)*, 14, 21

National Council of Book Development Board *(Pustak Vikas Mandal Ki-Rashtriya Parishad)*, 10

National Council of Educational Research and Training (NCERT) *(Rashtriya-Anusandhan Tatha Prashikshan-Parishad)*, 10, 26, 72

National Council of Teachers' Education *(Rashtriya Adhyapak Shiksha-Parishad)*, 10

National Council for Women's Education *(Rashtriya Mahila Shiksha-Parishad)*, 10

National Dairy Research Institute (NDRI) *(Rashtriya Dugdhashala Anusandhan-Sansthan)*, 11, 22, 64

National Demonstration *(Rashtrya Pradarshan)*, 10

National Institute of Basic Education *(Rashtriya Buniyadi Shiksha-Sansthan)*, 14

National Institute of Rural Development *(Rashtriya Gramin Vikash-Sansthan)*, 66

National Review Committee on Higher Secondary Education *(Rashtriya-Uchchatar)*, 26

National Service Scheme *(Rashtriya Seva Yojana)*, 23, 74

Third Five-year Plan *(Tirtiya Panch Barshiya Yojna)*, 21

training of development staff *(vikas karmachari prashikshan)*, 61

training of farmers *(Kisano ka prashikshan)*, 67

transfer of technology *(praudyogiki ka hastantaran)*, 10

Union Ministry of Agriculture and Irrigation *(Kendriya Krishi Aur-Sichaya Mantralya)*, 10, 23, 48

Union Ministry of Education *(Kendriya Shiksha Mantralaya)*, 17

University Education Commission (UEC) *(Vishwavidyalaya Shiksha Ayog)*, 17

University extension function *(vishwa vidyalaya prasar karya)*, 59

University Grants Commission (UGC) *(Vishwa Vidyalaya Anudan Ayog)*, 10, 23

university of agriculture and technology *(krishi tatha praudyogiki-Vishwavidyalaya)*, 17

upper primary *(uchcha prathmik)*, 11

Van-Mahotsava *(Tree-plantation Yojana)*, 25

vernacular middle school *(prakrit madhyamic pathshala)*, 10

vernacular training course *(prakrit prashikshan adhyayan)*, 10

veterinary assistant *(pashu chikitsa sahayak)*, 11

veterinary compounders *(pashu chikitsa compounder)*, 16, 31

veterinary education *(pashu chikitsa shiksha)*, 12

veterinary field assistant *(pashu chikitsa kshetra sahayak)*, 31

veterinary school *(pashu chikitsa vidyalaya)*, 11

veterinary teachers *(pashu chikitsa shikshak)*, 19

village-level worker *(gram sevak)*, 15

vocational agriculture *(vyavasayik krishi)*, 11

vocational courses *(vyavsayik adhyayan)*, 15

vocational education *(vyavsayik shiksha)*, 10, 15

vocational fisheries *(vyavsayik matasya)*, 15

vocational school *(vyavasayik vidyalaya)*, 11

vocational training *(vyavsayik prashikshan)*, 29

work experience *(byabsaik shiksha)*, 15

Selected references

ATWAL, A. S. (ed.). *New Concepts in Agricultural Education*. Punjab Agricultural University Press, 1969.

Education in India, 1972-73. New Delhi, Ministry of Education and Social Welfare, Government of India, 1978.

Education in India, 1974-76. New Delhi, Ministry of Education and Social Welfare, Government of India, 1977.

Fisheries Education and Training in India. Central Institute of Fisheries, Nautical and Engineering Training, Ministry of Agriculture and Irrigation, Government of India, 1977.

GAUTAM, O. P. *Growth of Agricultural Education*. New Delhi, Indian Council of Agricultural Research, 1972.

Handbook on Dairy Training in India, Karnal, National Dairy Research Institute, 1961.

I.C.A.R. Handbook. New Delhi, Indian Council of Agricultural Research, 1971.

The Indian Year Book on Education, 1961. New Delhi, National Council of Educational Research and Training, 1965.

Krishi Vigyan Kendra. An Innovative Institution. New Delhi, Indian Council of Agricultural Research, 1977.

NAIK, K. C. *Agricultural Education in India: Institutions and Organizations*, New Delhi, Indian Council of Agricultural Research, 1961.

——. *A History of Agricultural Universities*. A Committee on Institutional Co-operation, U.S. Agency for International Development Project, 1968.

PRASAD, C. *Farmers' Functional Literacy*. Literacy Discussion, Vol. VI, No. 3. Iran, International Institute of Adult Literacy Methods, 1975.

Punjab Agricultural Handbook. Ludhiana, Punjab Agricultural University, 1975.

Report of the Education Commission, 1964-66. New Delhi, Government of India, Ministry of Education, 1966.

Report of the National Commission on Agriculture—Part IX, Forestry. New Delhi, Government of India, Ministry of Agriculture and Irrigation, 1976.

Report of the National Commission on Agriculture—Part XI, Research, Education and Extension. New Delhi, Government of India, Ministry of Agriculture and Irrigation, 1976.

Report of the Review Committee on Agricultural Universities. New Delhi, Indian Council of Agricultural Research, 1978.

Report of the Second Joint Indo-American Team on Agricultural Education, Research and Extension. New Delhi, ICAR, 1960.

Report of the University Education Commission, 1948-49. New Delhi, Government of India, 1949.

Resident Instruction Bulletin. Ludhiana, Punjab Agricultural University, 1974.

Training of Fisheries Operatives Personnel During the V Plan Period. New Delhi, Central Institute of Fisheries Operatives, Ministry of Agriculture and Irrigation, Government of India. (Mimeo.)

DATE DUE

ED. 80/XXVIII. 4/A